COMBUSTION AND INCINERATION PROCESSES

POLLUTION ENGINEERING AND TECHNOLOGY

A Series of Reference Books and Textbooks

EDITORS

RICHARD A. YOUNG

Editor, Pollution Engineering
Technical Publishing Company
Barrington, Illinois

PAUL N. CHEREMISINOFF

Associate Professor
of Environmental Engineering
New Jersey Institute of Technology
Newark, New Jersey

1. Energy from Solid Wastes, *Paul N. Cheremisinoff and Angelo C. Morresi*

2. Air Pollution Control and Design Handbook (in two parts), *edited by Paul N. Cheremisinoff and Richard A. Young*

3. Wastewater Renovation and Reuse, *edited by Frank M. D'Itri*

4. Water and Wastewater Treatment: Calculations for Chemical and Physical Processes, *Michael J. Humenick, Jr.*

5. Biofouling Control Procedures, *edited by Loren D. Jensen*

6. Managing the Heavy Metals on the Land, *G. W. Leeper*

7. Combustion and Incineration Processes: Applications in Environmental Engineering, *Walter R. Niessen*

Additional volumes in preparation

COMBUSTION AND INCINERATION PROCESSES

Applications in Environmental Engineering

WALTER R. NIESSEN

Vice President, Industrial Engineering Division
Camp Dresser & McKee Inc.
Boston, Massachusetts

MARCEL DEKKER, INC.   New York and Basel

Library of Congress Cataloging in Publication Data

Niessen, Walter R   (1938-   )
   Combustion and incineration processes.

   (Pollution engineering and technology ; v. 7)
   Includes index.
   1. Incineration. 2. Combustion. I. Title.
II. Series.
TD796.N53   628'.445   77-17932
ISBN 0-8247-6656-3

MARCEL DEKKER, INC.
270 Madison Avenue, New York, New York 10016

Current printing (last digit):
10 9 8 7 6 5 4 3 2 1

PRINTED IN THE UNITED STATES OF AMERICA

I dedicate this book to my wife Dorothy and to my daughters Heidi and April who raked and painted and mowed and missed me on many an afternoon and evening during its preparation.

Purification by fire is an ancient concept, its applications noted in the earliest chapters of recorded history. The art and the technology of combustion (incineration) and pyrolysis as applied to environmental engineering problems draws on this experience, as well as the results of sophisticated contemporary research. To many engineers, however, combustion systems still hold an unnecessary mystery, pose unnecessary questions, and generate unnecessary mental barriers to their full exploitation as tools to solve tough problems. This book was written in an earnest attempt to thin the clouds of mystery, answer many of the questions (those for which answers are available), and provide a clearer way for the engineer to analyze, evaluate, and design solutions to environmental problems based on combustion.

The book describes combustion and combustion systems from a process viewpoint in an attempt to develop fundamental understanding rather than present simplistic design equations or nomographs. In large part, this approach was selected because combustion systems are complex and not readily susceptible to "cook-book" design methods. Consequently, considerable space is devoted to the basics: describing the chemical and physical processes which control system behavior.

In an effort to make the book as comprehensive as possible, a large number of topics have been dealt with. Specialists in particular fields may perhaps feel that the subjects in which they are interested have received inadequate treatment. This may be resolved in part by exploring the noted references, an activity also recommended to the newcomer to the field.

The publication of this book appears timely since current trends in environmental awareness and regulatory controls will prompt increases in the use of combustion technology as the preferred or only solution. In light of escalating construction costs, the soaring expense and diminishing availability of fossil fuels used as auxiliary energy sources (or the growing value of recovered energy), and the ever more stringent regulatory insistence on high performance regarding combustion efficiency and/or air pollutant emissions, the "black box" approach is increasingly unacceptable to the designer and to the prospective owner.

This book was prepared to meet the needs of many: students; educators; researchers; practicing civil, sanitary, mechanical, and chemical engineers; and the owners and operators of combustion systems of all types—but particularly those dealing with environmental problems. To serve this diverse audience, considerable effort has been expended to provide reference data, correlations, numerical examples, and other aids to fuller understanding and use.

Last (but of the greatest significance to me, personally), the book was written because I find the study and application of combustion to be an exciting and mind stretching experience: ever fascinating in its blend of predictability with surprise (though sometimes, the surprises are cruel in their impact). Combustion processes are and will continue to be useful resources in solving many of the pressing environmental problems of modern civilization. I sincerely hope that my efforts to share both contemporary combustion technology and my sense of excitement in the field will assist in responding to these problems.

In the preparation of this book, I have drawn from a broad spectrum of the published literature and on the thoughts, insights, and efforts of colleagues with whom I have been associated throughout my professional career. I am particularly grateful for the many contributions of my past associates at Arthur D. Little, Inc. and at the Massachusetts Institute of Technology, whose inspiration and perspiration contributed greatly to the substance of the book. Also, the many discussions and exchanges with my fellow members of the Incinerator Division (now the Solid Waste Processing Division) of the American Society of Mechanical Engineers have been of great value.

I must specifically acknowledge Professor Hoyt C. Hottel of MIT who introduced me to combustion and inspired me with his brilliance, Mr. Robert E. Zinn of ADL who patiently coached and taught me as I entered the field of incineration, and Professor Adel F. Sarofim of MIT whose technical insights and personal encouragement have been a major force in my professional growth.

I would like to acknowledge the support given by Roy F. Weston Inc. and Camp Dresser & McKee Inc. in underwriting the typing of the text drafts and the preparation of the art work. Particularly, I would thank Louise Miller, Bonnie Anderson and Joan Buckley who struggled through the many pages of handwritten text and equations in producing the draft.

Walter R. Niessen

Chapter 1

INTRODUCTION

For many waste streams, combustion (incineration) is an attractive or necessary <u>processing</u> step. In some cases (e.g., fume incineration or incineration of essentially ash-free liquids or solids), combustion processes may be properly called <u>disposal</u>. For most solids and many liquids, incineration is only a processing step, and liquid or solid residues remain for subsequent disposal by landfill or other means.

Incineration of wastes offers the following advantages or potential advantages:

1. Volume reduction, especially for bulky solids with a high combustible content

2. Detoxification, especially for combustible carcinogens, pathologically contaminated material, toxic organic compounds, or biologically active materials which would affect sewage treatment plants

3. Environmental impact mitigation, especially for organic materials which would leach from landfills or create odor nuisances

4. Regulatory compliance, especially for fumes containing odorous compounds, photoreactive organics, carbon monoxide, or other combustible materials subject to regulatory emission limitations

5. Energy recovery, especially when large quantities of waste are available and reliable markets for by-product fuel or steam are nearby

These advantages have justified development of a variety of incineration systems, of widely different complexity and function to meet the needs of municipalities and commercial and industrial firms and institutions.

Operating counter to these advantages are the following disadvantages of incineration:

1. Cost. In most instances, incineration is a costly waste processing step, both in initial investment and in operation.

2. Operating problems. Variability in waste composition and the severity of the incinerator environment result in many practical waste handling problems, high maintenance requirements, and equipment unreliability.

3. Staffing problems. The low status often accorded to waste disposal makes it difficult to obtain and retain qualified supervisory and operating staff.

4. Secondary environmental impacts. Many waste combustion systems result in the emission of odors, sulfur dioxide, hydrogen chloride, carbon monoxide, carcinogenic polynuclear hydrocarbons, nitrogen oxides, fly ash and particulate fumes, and other toxic or noxious materials into the atmosphere. Wastewaters from residue quenching or scrubber type air pollution control are often highly acidic and contain high levels of dissolved solids, abrasive suspended solids, biological and chemical oxygen demand, heavy metals, and pathogenic organisms. Lastly, residue disposal can present a variety of aesthetic, water pollution, and health-related problems.

5. Public sector reaction. Few incinerators are installed without arousing concern, close scrutiny, and, at times, hostility from the public at large and/or regulatory agencies.

6. Technical risk. Since changes in waste character are common (e.g., due to seasonal changes in municipal waste and process changes in industrial waste) and process analysis is difficult, a definite risk exists that a new incinerator may not work well or, in extreme cases, at all. Generally, changes in waste character invalidate any performance guarantee given by equipment vendors.

With all these disadvantages, incineration has persisted as an important concept in waste management. Indeed, some 20% of municipal solid waste and a somewhat smaller fraction of commercial and industrial waste was incinerated in 1970. In the late 1960s and early 1970s, the construction of new incineration units declined greatly, responding to the combined pressures of stronger air pollution control regulations and anti-incineration social pressure from the advocates of recycling. However, increasing concern over leachate generation and control (with consequent impacts on landfill cost) and the great increases in the value of energy occasioned by the energy crisis of the mid-1970s may well expand the role of incineration in the future.

Combustion processes are complicated. An analytical description of combustion system behavior requires consideration of:

1. Chemical reaction kinetics and equilibrium under nonisothermal, nonhomogeneous, nonsteady conditions

2. Fluid mechanics in nonisothermal, nonhomogeneous, reacting mixtures with heat release which can involve laminar, transition and turbulent, plug, recirculating, and swirling flows within geometrically complex enclosures

3. Heat transfer by conduction, convection, and radiation between gas volumes, liquids, and solids with high heat release rates and (with boiler systems) high heat withdrawal rates

In incineration applications, this complexity is often increased by frequent unpredictable shifts in fuel composition resulting in changes in heat release rate, and combustion characteristics (ignition temperature, air requirement, etc.). Compounding these process-related facets of waste combustion are the practical design and operating problems in materials handling, corrosion, odor, vector and vermin control, residue disposal, associated air and water pollution control, and myriad social, political, and regulatory pressures and constraints.

With these technical challenges facing the waste disposal technologist, it is a wonder that the state of the art has advanced beyond simple, batch-fed, refractory hearth systems. Indeed, incineration technology is still regarded by many as an art, too complex to understand.

The origins of such technical pessimism have arisen from many facts and practical realities:

1. Waste management has never before represented enough of a business opportunity to support extensive internal or sponsored research by equipment vendors, universities, or research institutions.

2. Municipal governments and most industries have had neither budgets nor inclination to fund extensive analysis efforts as part of the design process.

3. The technical complexity and "crudeness" of incineration has failed to stimulate investigations by the academic community.

4. The technical responsibility for waste disposal has usually been given to firms and individuals skilled in the civil and sanitary engineering disciplines, fields where high temperature, reacting, mixing, radiating (etc.) processes are not part of the standard curriculum.

In the author's opinion, such pessimism is extreme. To be sure, the physical situation is complex, but, drawing on the extensive scientific and

engineering literature in conventional combustion, the problems can be
made tractable. The remainder of this volume attempts to bring a measure
of structure and understanding to those wishing to analyze, design, and
operate incineration systems. Although the result cannot be expected to
answer all questions and anticipate all problems, it will give the student or
practicing engineer the quantitative and qualitative guidance and understanding to cope with this important sector of environmental control engineering.

The analytical methods and computational tools used draw heavily on
the disciplines of chemical and, to a lesser extent, mechanical engineering.
Recognizing that many readers may not be familiar with the terms and concepts involved, the early chapters review the fundamental analysis methods
of process engineering. Combustion and pyrolysis processes are then discussed, followed by a quantitative and qualitative review of the heat and
fluid mechanics aspects of combustion systems.

Building on the basic framework of combustion technology, combustion-
based waste disposal is then introduced: waste characterization, incinerator
systems, design principles and calculations.

Chapter 2

STOICHIOMETRY

Stoichiometry is the discipline of tracking matter (particularly the elements, partitioned in accord with the laws of chemical combining weights and proportions) and energy (in all its forms) in either batch or continuous processes. Since these quantities are conserved in the course of any process, the engineer can apply these principles of conservation to follow the course of combustion and flow processes, and compute air requirements, flow volumes, and velocities, temperatures, and other useful quantities. As a refinement, the engineer should acknowledge the fact that some reactions and heat transfer processes sometimes stop somewhat short of "completion" as a consequence of equilibrium limitations. Also, for some situations, the chemical reaction rate may limit the degree of completeness, especially when system residence time is short.

A. UNITS AND FUNDAMENTAL RELATIONSHIPS

1. Units

In analyzing combustion problems, it is advantageous to use the kilogram mole (or kilogram atom), the molecular (atomic) weight expressed in kilograms, as the unit quantity. The advantage derives from the facts that one molecular (atomic) weight of any compound (element) contains the same number of molecules (atoms) and that each mole of gas, at standard pressure and temperature, occupies the same volume. The volumetric identity holds if the gases are "ideal," an assumption which is acceptably accurate for gases at combustion temperatures.

2. Gas Laws

The behavior of "ideal gases" is described by the equation

$$PV = nRT \tag{1}$$

In this relationship P is the absolute pressure of the gas, V its volume, n the number of moles of gas, R the gas constant, and T the __absolute__ tempera- ture. Note that 273 (approximately) must be added to the Celsius tempera- ture and 460 to the Fahrenheit temperature to get the absolute temperature in degrees Kelvin(K) or degrees Rankine (°R), respectively.

The gas constant R applies to all gases but care must be given to as- sure compatibility of the units of R with those used for P, V, n, and T. Commonly used values of R are given in Table 1.

__Example 1.__ 10,000 kg/day of a spent absorbant containing 92% carbon, 6% ash, and 2% moisture is to be burned completely to generate carbon dioxide for process use. The exit temperature of the incinerator is 1000°C. How many kilogram moles and how many kilograms of CO_2 will be formed per minute? How many cubic meters per minute at a pressure of 1.04 atmos- pheres?

One must first determine the number of kilogram atoms per minute of carbon (atomic weight = 12) flowing in the waste feed: $0.92 \times 10,000/12 \times 24 \times 60 = 0.532$. Noting that with complete combustion each atom of carbon yields one molecule of carbon dioxide, the generation rate of CO_2 is also

TABLE 1

Values of the Gas Constant R for Ideal Gases

Energy	Pressure (P)	Volume (V)	Moles (n)	Temperature (T)	Gas Constant (R)
—	atm	m^3	kg mol	K	$0.08206 \ \dfrac{m^3 \ atm}{kg \ mol \ K}$
—	psia	ft^3	lb mol	°R	$1543 \ \dfrac{ft \ lb}{lb \ mol \ °R}$
—	atm	ft^3	lb mol	°R	$0.729 \ \dfrac{ft^3 \ atm}{lb \ mol \ °R}$
kcal	—	—	kg mol	K	$1.986 \ \dfrac{kcal}{kg \ mol \ K}$

0.532 kg mol/min. The weight flow will be $(0.532)(44) = 23.4$ kg/min of CO_2 (molecular weight = 44). T, in degrees Kelvin, is $1000 + 273 = 1273K$. Then from $PV = nRT$,

$$V = \frac{nRT}{P} = 0.532 \times \frac{0.08206 \times 1273}{1.04}$$

$$= 53.5 \text{ m}^3/\text{min } CO_2$$

Commonly in combustion calculations, one knows the number of moles and the temperature and it is desired to compute the volume. For these calculations, it is often convenient to obtain the answer by multiplying the volume at some reference or standard condition to the conditions of interest. For such calculations one uses the gas laws in terms of the molecular volume: 22.4 m^3 (359.3 ft^3) is the volume of 1 kg mol (lb mol) of an ideal gas under the standard conditions of 0°C or 273K and 1 atm (32°F or 492°R, 1 atm). Denoting the molecular volume as V_0, the gas law then yields

$$V = n \times V_0 \times \frac{P_0}{P} \times \frac{T}{T_0} \tag{2}$$

where P_0 and T_0 may be expressed in any consistent absolute units. For example, the gas volume from the preceding example could be calculated as

$$V = 0.532 \times 22.4 \times \frac{1.00}{1.04} \times \frac{1273}{273}$$

$$= 53.5 \text{ m}^3$$

It should be cautioned that the meaning of the term "standard conditions" may differ in the literature. For example, V_0 (in m^3 for kg mol and ft^3 for lb mol) for fan manufacturers and the natural gas industry is based on 60°F (15.6°C), and 1 atm (29.92 in. Hg or 14.7 lb/in.^{2} abs) where $V_0 = 23.7$ m^3 or 380 ft^3; while the manufactured gas industry uses 60°F, saturated with water vapor at 30 in. Hg abs for marketing ($V_0 = 24.0$ m^3 or 385 ft^3) but 60°F dry, at 1 atm ($V_0 = 23.6$ m^3 or 378 ft^3) for calculations.

Another useful identity derived from the ideal gas law, Dalton's law of partial pressures, equates the volume percent with the mole percent and partial pressure:

Volume percent = 100 (mole fraction)

$$= 100 \frac{\text{partial pressure}}{\text{total pressure}} \tag{3}$$

3. Energy

In this chapter, the basic unit of heat energy will be the kilogram calorie (kcal), the amount of heat necessary to raise the temperature of 1 kg of water 1°C. The Btu, the English system equivalent, relates to the energy to raise the temperature of 1 lb of water 1°F. Useful conversion factors are 1 kcal = 3.968 Btu and 1 kcal/kg = 1.8 Btu/lb.

a. Heat of Reaction

In many combustion calculations it is necessary to calculate the total heat release from the reported (or measured) heat of combustion of the combustible. As an example, for the complete combustion of methane, the reaction is

$$CH_4 \text{ (g)} + 2O_2 \text{ (g)} \longrightarrow CO_2 \text{ (g)} + 2H_2O \text{ (l)} + 212,950 \text{ kcal}$$

This indicates that to burn 1 kg mol of methane, 2 kg mol of oxygen are required. 1 kg mol of carbon dioxide and 2 kg mol of water result, and 212,950 kcal of heat are released. The subscripts g, l, and s denote the gaseous, liquid, and solid states of reactants or products. A pressure of 1 atm and an initial and final temperature of the reactants and products of 298K (20°C) are assumed, unless otherwise indicated. The heat of combustion given in this manner (with the water condensed) is known as the Higher Heating Value (HHV) and is the common way to report such data. Clearly, however, in the furnace, the sensible heat found in the flue gases will be lower than this by the latent heat of vaporization of the water (10,520 kcal/kg mol at 20°C) or 21,040 kcal/kg mol of methane. Thus the so-called Lower Heating Value (LHV) is 191,910 kcal corresponding to:

$$CH_4 \text{ (g)} + 2O_2 \text{ (g)} \longrightarrow CO_2 \text{ (g)} + 2H_2O \text{ (g)} + 191,910 \text{ kcal}$$

b. Sensible Heat of Gases

In many combustion calculations it is necessary to calculate the sensible heat content (enthalpy) of gases at elevated temperatures, or the change in enthalpy between two temperatures. To make such calculations it is useful to draw on the approximation that Mc_p, the molal specific heat at constant pressure (kcal/kg mol °C) is a function of temperature only, and independent of pressure and concentration; asymptotically approaching a value Mc_p° as the pressure approaches zero.

The enthalpy change (Δh) between temperature limits T_1 and T_2 is then given by

$$\Delta h = \int_{T_1}^{T_2} Mc_p^\circ \, dT \text{ kcal/kg mol} \tag{4}$$

This calculation may be carried out using an analytical relationship for $Mc_p^\circ(T)$ with the constants shown in Table 2 or, more simply, using a graphical presentation (Fig. 1) of the average molal heat capacity between a reference temperature of 15°C (60°F) and the abscissa temperatures. The

TABLE 2

Constants in Molal Heat Capacity (Mc_p°) Relationship with Temperature

$$(Mc_p^\circ = a + bT + cT^2)^a$$

Compound	a	b	c
H_2	6.92	0.153×10^{-3}	$+0.279 \times 10^{-6}$
O_2	6.95	2.326×10^{-3}	-0.770×10^{-6}
N_2	6.77	1.631×10^{-3}	-0.345×10^{-6}
Air	6.81	1.777×10^{-3}	-0.434×10^{-6}
CO	6.79	1.840×10^{-3}	-0.459×10^{-6}
CO_2	9.00	7.183×10^{-3}	-2.475×10^{-6}
H_2O	7.76	3.096×10^{-3}	-0.343×10^{-6}
NO	6.83	2.102×10^{-3}	-0.612×10^{-6}
SO_2	9.29	9.334×10^{-3}	-6.38×10^{-6}
HCl	6.45	1.975×10^{-3}	-0.547×10^{-6}

^{a} Where T is in °C and Mc_p° is in kcal/kg mol °C. Accuracy ~ 2%, 20°C to 1700°C.

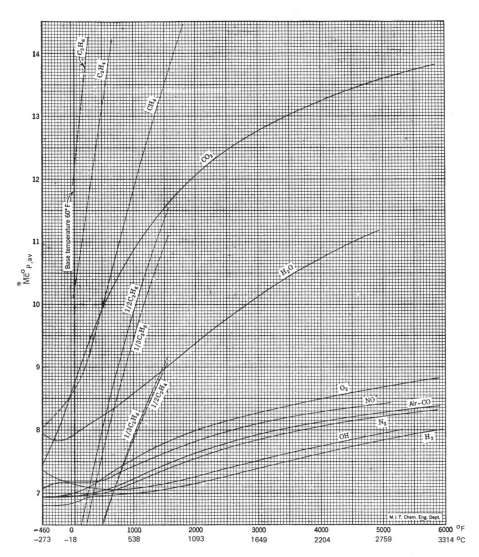

FIG. 1. Average molal heat capacity between 15°C and upper temperature. (Courtesy, Professor H. C. Hottel, Chemical Engineering Department, M.I.T.)

average molal heat capacity is calculated and used as follows:

$$Mc^\circ_{p,av} = \frac{\int_{15}^{T} Mc^\circ_p \, dT}{T - 15} \tag{5}$$

$$\Delta h^T_{15} = nMc^\circ_{p,av}(T - 15) \text{ kcal} \tag{6}$$

Example 2. What is the sensible heat content of 68 kg of carbon dioxide at 1200°C relative to 15°C, and how much heat must be extracted to bring the temperature to 300°C?

First, determine that the number of moles of $CO_2 = 68/44 = 1.55$. From Table 2 the heat content at 1200°C is given by:

$$H = 1.55 \int_{15}^{1200} (9.00 + 7.183 \times 10^{-3}T - 2.475 \times 10^{-6}T^2) \, dt$$

$$= \left[1.55\left(9T + \frac{7.183 \times 10^{-3}T^2}{2} - \frac{2.475 \times 10^{-6}T^3}{3}\right)\right]_{15}^{1200}$$

$$= 22,340 \text{ kcal}$$

A similar calculation for an upper limit of 300°C yields 4370 kcal and a net heat extraction requirement of 22,340 - 4370 = 17,970 kcal. Alternatively, using Fig. 1, the heat content at

$$1200°C = 1.55(12.3)(1200 - 15) = 22,600 \text{ kcal}$$

$$\text{and at } 300°C = 1.55(10.1)(300 - 15) = 4,460 \text{ kcal}$$

and the approximate heat loss,

$$1200 \text{ to } 300°C = 18,140 \text{ kcal}$$

c. Latent Heat

The change in state of elements and compounds, for example, from solid (s) to liquid (l), or from liquid (l) to gas (g), is accompanied by a heat effect: the latent heat of fusion, sublimation, or vaporization for the state changes (s) ⟶ (l), (s) ⟶ (g), and (l) ⟶ (g), respectively. This heat effect is often considered negligible, as for the heat of vaporization of fuel oil. Particularly in incineration calculations (where a high waste moisture

TABLE 3

Latent Heat Effects for Changes in State of Common Materials

Material	State change	Temperature (°C)	Latent heat of indicated state change (kcal/kg)	(kcal/kg mol)	(Btu/lb)
Water	Fusion	0	80	1435	144
Water	Vaporization	100	540	9712	971
Acetone	Vaporization	56	125	7221	224
Benzene	Vaporization	80	94	7355	170
i-Butyl alcohol	Vaporization	107	138	12,420	248
n-Decane	Vaporization	160	60	8548	108
Methanol	Vaporization	65	263	12,614	473
Turpentine	Vaporization	156	69	9330	123
Zinc	Fusion	419	28	1839	51

content is common), these effects can be very significant. For reference, several values are given in Table 3.

4. Basis of Computation

In order to be clear and accurate in combustion calculations it is important, as the first step in detailed problem analysis, to identify the thing about which one is talking. In this chapter, the term basis will be used. In the course of prolonged analyses, it may become useful to shift bases although the advantages are often countered by the lack of strict comparability between intermediate and final results.

As the initial step, therefore, the analyst should choose and write down the basis: a given weight of one feed material (e.g., 100 kg of waste) or element, or a unit time of operation. The latter is usually equivalent to a weight, however, and in general the weight basis is preferred.

5. Approach to Computation

Although the skilled analyst may elect to skip one or more steps because of limited data or lack of utility, the following sequence of steps are strongly recommended:

1. Sketch a flow sheet. Indicate all flows of heat and material, including recirculation streams. Document all basic data on the sketch including heat effects.

2. Select a basis and annotate the sketch to show all known flows of heat or material relative to that basis.

3. Apply material, elemental, and component balances.

4. Use energy balances.

5. Apply known equilibrium relationships.

6. Apply known reaction rate relationships.

7. Review the previous steps incorporating the refinements from subsequent stages into the simpler, earlier work.

B. MATERIAL BALANCES

A material balance is a quantitative expression of the law of conservation of matter: what goes in comes out (unless it stays behind).

Input = output + accumulation

This expression is always true for <u>elements</u> flowing through combustion systems (except for minor deviations in the unusual case where radioactive materials are involved). It is often not true for <u>compounds</u> participating in combustion reactions.

The basic data used in material balances can include analyses (fuel, waste, residue, gases in the system) coupled with some rate data (usually feed rate). Coupled with these data are fundamental relationships which prescribe the combining proportions in molecules (e.g., two atoms of oxygen with one of carbon in one molecule of carbon dioxide), and those which indicate the course and heat effect of chemical reactions.

1. Analyses

Several types of analysis are available to the analyst. They are briefly reviewed here.

a. Ultimate Analysis

For fuels, the term <u>ultimate analysis</u> refers to an analysis routine which includes: moisture (loss in weight, for solid fuels, at 105°C), combined water (equivalent to the oxygen), carbon, available or net hydrogen (that other than in moisture and combined water), sulfur, nitrogen, and "ash." In waste disposal calculations it may be appropriate to request analyses for chlorine, metals (such as lead) which have oxides which are in the vapor state at combustion temperatures, and other elements which could influence the combustion process or would be important to air pollution or water pollution aspects. It should be noted that oxygen is often determined by difference, and thus the "combined water" value may be in error. Also note that the reported weight of ash will be increased by the oxidation of metals in the sample or decreased by the release of CO_2 from carbonates, by loss of water of hydration, or by oxidation of sulfides.

Concern should also be given to the reported moisture level to assure that it is representative of the material "as-fired." Often, wastes are supplied to the laboratory after air drying either because the sampling team decided (without consultation) that such a step would be "good" or because insufficient attention was given to postsampling moisture loss. Not uncommonly, a waste sample may not be representative because the sampling team wanted to give you the "best they could find" or because they did not wish to handle some undesirable (e.g., garbage) or awkward (e.g., a large pallet) waste components. Let us leave this topic with an injunction to the engineer: <u>know the sampling, sample conditioning, and laboratory methods before trusting the data.</u>

b. Proximate Analysis

As will be seen, moisture content and combustible volatilization characteristics are important properties affecting combustion requirements and pyrolysis behavior of solid wastes and fuels. A relatively low cost laboratory test which reports these properties is called the <u>proximate analysis</u> [1]. The procedure includes the following steps:

1. Heat 1 hr at 104 to 110°C. Report weight loss as <u>moisture.</u>

2. Ignite in a covered crucible for 7 min at 950°C and report the weight loss (combined water, hydrogen, and the portion of the carbon initially present as or converted to volatile hydrocarbons) as <u>volatile combustible matter.</u>

3. Ignite in an open crucible at 725°C to constant weight and report weight loss as <u>fixed carbon.</u>

4. Report the residual mass as <u>ash.</u>

A useful application of the ultimate analysis is in the Dulong Formula, developed for use with coal but offering the opportunity to roughly estimate the heating value of a waste. The underlying assumption of the Dulong Formula [2] assumes a negligible heat of formation of the organic matter relative to the heat of combustion of the elements:

$$\text{kcal/kg (Dry basis)} = 8086 \text{ (wt fraction carbon)}$$
$$+ 34,488 \text{ (wt fraction hydrogen)}$$
$$- 0.125 \text{ (wt fraction oxygen)}$$
$$+ 2252 \text{ (wt fraction sulfur)} \qquad (7)$$

A more complete expression [3], which compensates for the heat losses in calcining carbonate carbon and for nitrogen requires knowledge of the weight fraction of organic and inorganic (carbonate) carbon, oxygen, sulfur, hydrogen, and nitrogen (denoted C_{org}, C_{inorg}, O, S, H, and N, respectively) in the waste sample. The equation is

$$\text{kcal/kg (Dry basis)} = 7,831 \, C_{org} + 35,932\left(H - \frac{O}{8}\right)$$
$$+ 2,212 \, S - 3,545 \, C_{inorg} + 1,187 \, O + 578 \, N \qquad (8)$$

Example 3. Estimate the heating value of a refuse with the composition: 45.85% organic carbon, 0.83% inorganic carbon, 6.61% hydrogen, 35.94% oxygen, 1.03% nitrogen, 0.1% sulfur, and 9.64% ash.

$$\text{kcal/kg (Dry basis)} = (7.831)(0.4585) + (35,932)\left(0.0661 - \frac{0.3594}{8}\right)$$
$$+ (2,212)(0.001) - (3,545)(0.0083) + (1,187)(0.3594)$$
$$+ (578)(0.0103)$$
$$= 4,836 \text{ kcal/kg}$$

c. Orsat Analysis

The Orsat analysis is, perhaps, the easiest and most useful test of combustor performance. Indeed, it is appropriate to periodically confirm the accuracy of more sophisticated instrumentation using the Orsat apparatus.

In use, a measured volume of flue gas is drawn into the apparatus and allowed to equilibrate with room temperature and to become saturated with water vapor. (Even if mercury is used as the confining fluid in the gas measuring burette, a drop of water should be maintained on top of the mercury column.) By raising or lowering the vessel of confining fluid, the gases are then repeatedly bubbled through selective absorbing fluids until a constant volume reading is obtained. Relative volume changes are then reported as the percent, on a dry basis of carbon dioxide, carbon monoxide, and oxygen. Nitrogen is reported by difference.

As a cautionary statement, it should be noted that sulfur dioxide will often be reported as carbon dioxide since absorption of both gases will occur in the alkaline medium used. Also, both gas percentages can be appreciably in error if water (instead of mercury) is used as the confining medium, due to the solution of the gases (even if the water is made acid and is virtually saturated with salt). Lastly, far too often, the absorbent (particularly the pyrogallate used for oxygen absorption) becomes deactivated through age, overuse, or improper storage. The resulting limited absorbing capacity leads the analyst to report inaccurately.

d. Waste Analysis

Incineration systems, unlike more conventional combustors, are often charged with materials of compositions that vary widely over time and which are highly complex mixtures of waste streams, off-specification products, plant trash, and so forth. The analysis of these wastes must often be a compromise.

In residential waste incineration, for example, what is a shoe: (1) a shoe? (2) 0.5 kg in the "leather and rubber" category? (3) 0.2 kg leather, 0.18 kg rubber, 0.02 kg iron nails, etc.? (4) 8% moisture, 71% combustible, 21% residue, heating value of 3800 kcal/kg? (5) the composition as given by ultimate analysis? or (6) properties as given by a proximate analysis? These are the questions the waste engineer must ponder as they impinge upon the adequateness of his design, the need for rigorous detail (in consideration of feed variability), and, importantly, the sampling and analysis budget allocation.

The final decision should be based on the impact of errors on:

Materials handling: bulk density, storability, explosion and fire hazard, etc.

Fan requirements: combustion air and draft fans

Heat release rate: per square meter, per cubic meter of the combustor

Materials problems: refractory or fireside boiler tube attack, corrosion in tanks, pipes, or storage bins, etc.

Secondary environmental problems: air, water, and residue-related pollution

Process economics: heat recovery rate, labor requirements, utility usage, etc.

There are no simple rules in this matter. Judgments are necessary on a case-by-case basis. The techniques described below, however, give the analyst tools to explore many of these effects on paper, at a cost much lower than detailed field testing and laboratory analysis and far less than is incurred after an incineration furnace has been installed and fails to operate satisfactorily.

2. Balances Based on Fuel Analysis

Balances on elements in the fuel or waste allow one to calculate the amount of air theoretically required to completely oxidize the carbon, net hydrogen, sulfur, etc. This quantity of air (known as the <u>theoretical</u> or <u>stoichiometric</u> air requirement) is often insufficient in a practical combustor and <u>excess air,</u> defined as a percentage of the stoichiometric air quantity, is usually supplied. For example, an operation at 50% excess air denotes combustion where 1.5 times the stoichiometric air requirement has been supplied. Typical excess air levels for several fuels are given in Table 4.

At a specified excess air level, elemental balances allow computation of the flue gas composition. The method of analysis is best illustrated by an example:

<u>Example 4.</u> Calculate the air requirement and products of combustion when burning at 30% excess air, 75 kg/hr of a waste liquid having the ultimate analysis: 12.2% moisture, 71.0% carbon, 9.2% hydrogen, 3.4% sulfur, 2.1% oxygen, 0.6% nitrogen, and 1.5% ash. The combustion air is at 15.5°C, 70% relative humidity. The sequence of computations is shown in Table 5.

Several elements of the analysis should be noted.

Line 1. Carbon is assumed here to burn completely to carbon dioxide. In practice, some carbon may be incompletely burned, forming carbon monoxide, and some may end up as unburned carbon char in solid residues or as part of the particulate matter leaving (as soot or char fragments) in the effluent gas. The fraction of carbon as CO usually ranges from less than 1% in gaseous waste burners to as high as 2 to 4% in solids burners. Carbon as char in the residue can be as high as 10% to 15% of the carbon feed in solid waste burners.

Line 2. Hydrogen in the waste (other than the hydrogen in moisture) increases the amount of combustion air but does not appear in the Orsat analysis (Lines 16 to 18).

TABLE 4

Typical Excess Air and/or Maximum Percent CO_2 for Waste and Fossil Fuel Combustors

Fuel	Type of furnace or burner	Typical excess air (%)	Maximum % CO_2 (volume)[a]
Acid sludge	Cone and flat-flame type burners, steam atomized	10–15	—
Bagasse	All	20–35	19.4–20.5 (20.3)
Black liquor	Recovery furnaces	5–7	18.6 (without salt cake)
Blast furnace gas	Intertube nozzle-type burners	15–18	20.0–26.9 (24.7)
Carbon (pure)	—	—	20.9
Charcoal	—	—	18.6
Anthracite coal	—	—	19.3–20.0
Bituminous coal	Range for nongrate fired	—	17.7–19.3 (18.5)
Pulverized	Waterwall furnace, wet or dry bottom	15–20	—
	Partial waterwall, dry bottom	15–40	—

Crushed	Cyclone furnaces	10-15	—
Lump or crushed	Spreader stoker	30-60	—
	Vibrating grate (water cooled)	30-60	—
	Chain or traveling grate	15-50	—
	Underfire stoker	20-50	—
Coke oven gas	Multifuel burner	7-12	9.5-12.7 (11.1)
Lignite coke	—	—	19.2
Natural gas	Register-type burner	5-10	11.6-12.7 (12.2)
Oil (fuel)	Register-type	5-10	14.25-15.5
	Multifuel burners and flat flame	10-20	14.75-16.35
Refinery gas	Multifuel burner	7-12	—
Tar and pitch	—	—	17.5-18.4
Softwood	—	—	18.7-20.4
Hardwood	—	—	19.5-20.5

[a]Values in parentheses are typical.

TABLE 5

Calculations for Example 4[a]

Line	Component	kg	Atoms or moles[b]	Combustion product	Theoretical moles O_2 required
1	Carbon (C)	71.0	5.917	CO_2	5.917
2	Hydrogen (H_2)	9.2	4.600	H_2O	2.300
3	Sulfur (S)	3.4	0.106	SO_2	0.106
4	Oxygen (O_2)	2.1	0.065	—	(0.065)
5	Nitrogen (N_2)	0.6	0.021	N_2	0.0
6	Moisture (H_2O)	12.2	0.678	—	0.0
7	Ash	1.5	N/A	—	0.0
8	Total	100.0	11.387		8.258
9	Moles nitrogen in stoichiometric air[c]	(79/21)(8.258)			
10	Moles nitrogen in excess air	(0.3)(79/21)(8.258)			
11	Moles oxygen in excess air	(0.3)(8.258)			
12	Moles moisture in combustion air[d]				
13	Total moles in flue gas				
14	Volume (mole) percent in wet flue gas				
15	Orsat (dry) flue gas analysis—moles				
16	A. With selective SO_2 testing—vol %				
17	B. With alkaline CO_2 testing only—vol %				
18	C. With SO_2 loss in testing—vol %				

[a] Basis: 100 kg of waste.
[b] The symbol in the component column shows whether these are kg moles or kg atoms.
[c] Throughout this chapter, dry combustion air is assumed to contain 21.0% oxygen by volume and 79.0% nitrogen.
[d] Calculated as follows: $(0.008/18)[(31.077 + 9.323)(28) + (2.477)(32)]$ based on the assumption of 0.008 kg water vapor per kg bone-dry air; found from standard psychrometric charts.

Moles formed in stoichiometric combustion					
CO_2	H_2O	SO_2	N_2	O_2	Total
5.917	0.0	0.0	0.0	0.0	5.917
0.0	4.600	0.0	0.0	0.0	4.600
0.0	0.0	0.106	0.0	0.0	0.106
0.0	0.0	0.0	0.0	0.0	0.0
0.0	0.0	0.0	0.021	0.0	0.021
0.0	0.678	0.0	0.0	0.0	0.678
0.0	0.0	0.0	0.0	0.0	0.0
5.917	5.278	0.106	0.021	0	11.322
			31.077		31.077
			9.323		9.323
				2.477	2.477
	0.538				0.538
5.917	5.816	0.106	40.421	2.477	54.737
10.810	10.625	0.194	73.846	4.525	100.000
5.917		0.106	40.421	2.477	48.921
12.095	N/A	0.217	82.625	5.063	100.0
12.312	N/A	N/A	82.625	5.063	100.0
12.127	N/A	N/A	82.804	5.074	100.0

Line 3. Sulfur in the waste as sulfide or organic sulfur increases the amount of combustion air required in burning to SO_2. Inorganic sulfates may leave as ash or be reduced to SO_2. Usually, only 1 to 3% of the sulfur is found as SO_3. If SO_2 is not analyzed selectively (Line 17), it is usually reported out as carbon dioxide. If care is not taken to avoid solution of the SO_2 in the sampling and analysis steps, however, it acts to distort the $CO_2/N_2/O_2$ values. This is worthy of special consideration for waste disposal where the sulfur concentration may be high.

Line 4. Oxygen in the waste reduces the amount of required combustion air.

Line 12. Moisture entering with the combustion air can be seen to be small and is often neglected. As noted, psychrometric data relating the water content of atmospheric air to various conditions of temperature, relative humidity, etc., can be obtained from standard references (e.g., Ref. 4) and typical meteorological data for the region of interest from local weather stations or reference books (e.g., Ref. 5). Atmospheric moisture content should be considered in fan selection, and if the resulting flue gas moisture affects the process (e.g., in sulfuric acid manufacture by the Contact Process).

Although this problem considered only carbon, hydrogen, oxygen, nitrogen, and sulfur waste components and assumed single products of combustion, the analyst should review waste composition thoroughly and consider the range of possible secondary reactions if the flue gas analysis is especially important (e.g., in regard to secondary air pollution problems). A few of these concerns include:

Carbon Monoxide. As discussed later, CO may be formed in appreciable quantities in solids burning and in systems operated under near stoichiometric conditions.

Chlorine. Chlorine appearing in the waste as inorganic salts (e.g., NaCl, $CaCl_2$) will most likely remain as the salt. Organic chlorides, however, result primarily in the formation of hydrogen chloride (to the extent of the hydrogen available from all sources, including moisture in the combustion air; excess halogen appears as the element).

Metals. Metals usually burn to the oxide, although in burning solid wastes a large fraction of massive metal feed (e.g., tin cans, sheet steel, etc.) is unoxidized.

Thermal Decomposition. Some compounds in the feed may be decomposed at combustor temperatures. Carbonates, for example, may dissociate to form an oxide and CO_2. Ammonium compounds may dissociate, whereupon the ammonia burns and the anion (e.g., sulfate) appears as a particulate or gaseous component in the flue gas (e.g.,

SO_2 or SO_3 from sulfate). Sulfides may "roast" to form the oxide and release SO_2.

3. Balances Based on Flue Gas Analysis

In many instances, the analyst is called upon to evaluate an operating waste disposal system. In such studies, accurate data on the flue gas composition are readily obtainable and offer a low cost means to characterize the operation and the feed waste.

One important combustor and combustion characteristic which can be immediately computed from the Orsat (dry) flue gas analysis is the percentage excess air:

$$\text{Percentage excess air} = \frac{[O_2 - 0.5(CO + H_2)]100}{0.266\ N_2 - O_2 + 0.5(CO + H_2)} \qquad (9a)$$

where O_2, N_2, etc., are the volume percentages of the gases. If only the CO_2 content of the flue gas is known for a given fuel where the concentration of CO_2 at zero excess air is CO_2^*, the percentage excess air is given by:

$$\text{Percentage excess air} = \frac{7900(CO_2^* - CO_2)}{CO_2(100 - CO_2^*)} \qquad (9b)$$

Example 5. The flue gas from a waste incinerator burning a waste believed to have little nitrogen or oxygen has an Orsat analysis (using alkaline CO_2 absorbent) of 12.3% CO_2, 5.1% O_2, and the rest nitrogen and inerts. From these data calculate:

1. The weight ratio of hydrogen to carbon in the waste

2. The percent carbon and hydrogen in the dry waste

3. The kilograms dry air used per pound of dry waste

4. The percent excess air used

5. The moles of exhaust gas discharged from the unit per kilogram dry waste burned

Note that this example is derived from Example 4. From comparison, the "actual" values based on the prior example are given in parentheses.

---

*See Table 4.

Basis: 100 mol of dry exhaust gas

Component	Moles	Moles O_2
CO_2 (+ SO_2)	12.3	12.3
O_2	5.1	5.1
N_2	82.6	—
Total	100.0	17.4

$$82.6(21/79) = 22.0 \text{ Accounted for from air}$$

Difference = 4.6 mol O_2 disappearance

H_2 burned	2(4.6) ≐ 9.2 mol	=	18.4 kg
C burned	12(12.3 mol)	=	147.6 kg
Total weight of waste			166.0 (vs. 179.2 kg)

1. Weight ratio of hydrogen to carbon: $(18.4/147.6) = 0.125$ (vs. 0.130)

2. Percent (by weight) C in dry fuel: $(147.6/166.0)(100) = 88.92$ (vs. 80.86% for carbon, 84.74% for total of carbon and sulfur)

3. Kg dry air per kg of dry waste: First, calculate the weight of air resulting in 1 mol of dry exhaust gas from a nitrogen balance:

$$\frac{1}{100} (82.6 \text{ mol } N_2)\left(\frac{1}{0.79} \text{ moles } N_2 \text{ per mole air}\right)(29 \text{ kg air/mol})$$

$$= 20.32 \text{ kg air/mol dry exhaust gas}$$

then

$$30.32 \left(\frac{1}{1.66} \text{ mol dry exhaust gas per kg waste}\right)$$

$$= 18.27 \text{ kg dry air/kg dry waste (vs. 16.80)}$$

4. Percent excess air:

The oxygen <u>necessary</u>* for combustion is $12.3 + 4.6 = 16.9$ mol. The oxygen <u>unnecessary</u>* for combustion = 5.1 mol. The <u>total</u> oxygen = 22.0 mol.

---

*Note that the <u>necessary</u> oxygen would have to be increased and the <u>unnecessary</u> oxygen decreased if incompletely burned components (such as CO) were present.

The percent excess air (or oxygen) may be calculated as:

$$(100)\frac{\text{unnecessary}}{\text{total—unnecessary}} = \frac{100(5.1)}{22.0 - 5.1} \qquad = 30.2\% \text{ (vs. } 30\%) \quad (10a)$$

$$(100)\frac{\text{unnecessary}}{\text{necessary}} \qquad = \frac{100(5.1)}{16.9} \qquad = 30.2\% \qquad\qquad (10b)$$

$$(100)\frac{\text{total—necessary}}{\text{necessary}} = \frac{100(22 - 16.9)}{16.9} = 30.2\% \qquad\qquad (10c)$$

5. Moles of exhaust gas per kg of dry waste, noting that 9.2 mol of water vapor must be added to the dry gas flow,

$$\frac{100 + 9.2}{166} = 0.658 \text{ mol/kg fuel}$$

(vs. 0.623 for dry waste and 0.547 for true waste)

Lessons to be learned from comparison of the results of Example 5 with the "true"situation from Example 4 are:

Waste analysis data are particularly important in calculating combustion air requirements for design.

Waste moisture data are important in determining flue gas rates (induced draft fan pollution control and stack sizing).

Nonetheless, considerable insight on the nature of the waste can be gained from stack gas analysis.

4. Cross-Checking Between Fuel and Flue Gas Analysis

In studying waste incineration processes, the skilled analyst should take all reasonable opportunities to assemble several independent data sets and, using the techniques illustrated above, cross-check for consistency. Then, when errors appear, application of insights as to probable sampling or analytical error and the potential for waste variability can allow convergence on a "best" solution. Frankly, stopping short of such cross-checking is foolish.

C. ENERGY BALANCES

An energy balance is a quantitative expression of the law of conservation of energy. In waste incineration system analysis, five energy quantities are of prime interest:

Chemical Energy: the heat of chemical reaction which includes combustion and dissociation

Latent Heat: the heat to change the state of materials, which includes sublimation, fusion, and vaporization

Sensible Heat: the heat content related to the temperature of materials

Heat Losses: to the walls of combustion systems

Useful Heat: delivered to boiler tubes or to materials being heated by the combustion process

For energy calculations, only a condition of materials at absolute zero (-273°C, -460°F) provides a truly basic energy reference. For convenience, however, a condition of 1 atm and 15.6°C (60°F) is often used for sensible heat content. Most laboratory data on chemical reaction and latent energy is reported for a reference temperature of 15 to 25°C. In the latter case, although theory would dictate that correction to a consistent temperature base is appropriate, the errors incurred are generally minor. Thus, tabulated values of heat of combustion and latent heat are ordinarily used without correction.

Typically, heats of combustion are determined experimentally using a bomb calorimeter. In this method, a small quantity (often about 1 g) of material is ignited and burned with sodium peroxide in a heavy-walled, sealed container immersed in water. The resulting heat release is determined from the final temperature rise of the bomb, water, and associated container. Three consequences arise:

The final temperature is in the range 20 to 30°C, such that the measured (and reported) heat effect includes the latent heat of condensation of the water vapor formed in combustion. This so-called higher heating value is thus substantially greater than the heat appearing in the flue gas stream as sensible heat. The values for which the water is assumed to be condensed (the lower heating value) is corrected for this important heat effect. Unless otherwise specified, most reported heat of combustion data are for the higher heating value.

The quantity of waste tested is very small, thus greatly increasing the problem of obtaining a representative sample, especially for solid wastes. Unfortunately, larger combustion bombs are highly impractical, so this problem will continue to persist. One solution, having applicability in many situations, is to analyze the waste to identify its major constituents and then "construct" the heat content of the mixed waste as a weighted average.

The nature of the ignition process in the combustion bomb makes tests of liquids or moist sludges very difficult.

Table 6 and Appendix C give values for the heat of combustion of many common materials.

The uniformity of the higher heating value (HHV) per mole of oxygen at a stoichiometric mixture is worthy of note, particularly for the hydrocarbons where the average is about 110,000 kcal/mol O_2. This uniformity can be useful in estimating the average heat of combustion of incinerated wastes based on the air requirement determined in operating equipment.

Example 6. A waste sludge is to be burned in a well-insulated combustion chamber followed by a waste heat boiler. The sludge has the following composition:

Component	Weight percent
Benzene	70.0
Carbon (coke)	8.2
Ash	9.3
Moisture	12.5

The initial concept proposed is to burn the waste using only 10% excess air (available preheated to 300°C) in order to maximize the heat recovered from the flue gases. Assuming 5% heat loss from both the combustion chamber and the boiler and neglecting dissociation, at what temperature will the flue gases enter the boiler? At a burning rate of 1100 kg/hr, what will be the steaming rate if the stack temperature is 180°C (about 350°F) and the boiler steam is saturated at 15.8 atm? Boiler feedwater is the saturated liquid at 100°C.

A diagram of the process is given in Fig. 2.

Let the datum temperature be 15.5°C (60°F) and the basis be 100 kg of waste. The chemical heat input is given by:

Component	kg	Moles	HHV (kcal)	Total heat input (kg cal)	Required mol O_2
Benzene	70.0	0.897	787,200	706,450	6.73
Carbon	8.2	0.683	97,000	66,280	0.68
Ash	9.3	—	0	0	0.00
Moisture	12.5	0.694	0	0	0.00
Total	100.0			772,730	7.41

TABLE 6

Molal Heat of Combustion at Constant Pressure (kcal/kg mol)

Fuel	Higher heating value (HHV) (water condensed)	Lower heating value (LHV) (water uncondensed)	HHV per mole O_2	Temperature (°C)	Reference
Carbon (C to CO_2)	97,000	97,000	97,000	25	6
Hydrogen (H_2)	68,318	57,798	136,636	25	7
Carbon monoxide (CO)	67,636	67,636	135,272	25	7
Methane (CH_4)	212,950	191,910	106,475	18	8
Ethane (C_2H_6)	373,050	341,500	106,585	18	8
Ethylene (C_2H_4)	336,600	315,600	112,200	18	8
Acetylene (C_2H_2)	311,100	300,600	124,440	18	8
Propane (C_3H_8)	526,300	484,200	105,260	18–20	9
Propylene (C_3H_6)	490,200	458,700	108,933	18–20	9
Benzene (C_6H_6)	787,200	755,700	104,960	18–20	9
Sulfur (rhombic, to SO_2)	70,920	70,920	70,920	18	8

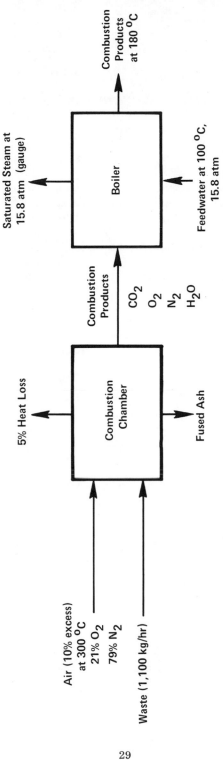

FIG. 2. Process diagram for Example 6.

The air requirement (at 10% excess) is calculated to be:

Total mol O_2 (7.41)(1.1)   = 8.15 mol

Total mol N_2 $(7.41)(1.1)\left(\dfrac{79}{21}\right)$ = $\underline{30.66}$

$\hphantom{Total mol N_2   (7.41)(1.1)(79/21) = }$ 38.81 mol

The air flow rate is, therefore, 918.8 sm^3 (standard cubic meters[*]) at 15.5°C or 1823.5 am^3 (actual cubic meters[†]) at 300°C. The sensible heat of the incoming air is given from (see Fig. 1 for heat capacity):

(38.81)(300 - 15.5)(7.08) = 78,170 kcal

Therefore, the total energy input is 772,730 + 78,170 = 850,900 kcal. The flue gas composition is given by:

Component	Required mol O_2	Mol CO_2	Mol H_2O	Mol O_2 (excess)	Mol N_2 (1.1)(79/21)(Req'd. O_2)
Benzene	6.73	5.382	2.691	0.673	27.849
Carbon	0.68	0.683	0.0	0.068	2.814
Ash	0.0	0.0	0.0	0.0	0.0
Moisture	0.0	0.0	0.694	0.0	0.0
Total	7.41	6.065	3.385	0.741	30.663
Volume percent (40.854 mol total)		14.846	8.286	1.814	75.055

To find the exit temperature of the combustion chamber and the steaming rate, it is useful to construct a plot of the heat content of the gas stream as a function of temperature.

The sensible and latent heat content of the gases and solid ash is computed as shown in Table 7. The results of these computations are plotted in Fig. 3. The flows of thermal energy are then:

---

[*]Volume (of a gas) at standard conditions of temperature and pressure.
[†]Volume (of a gas) at whatever temperature and pressure exist at the time.

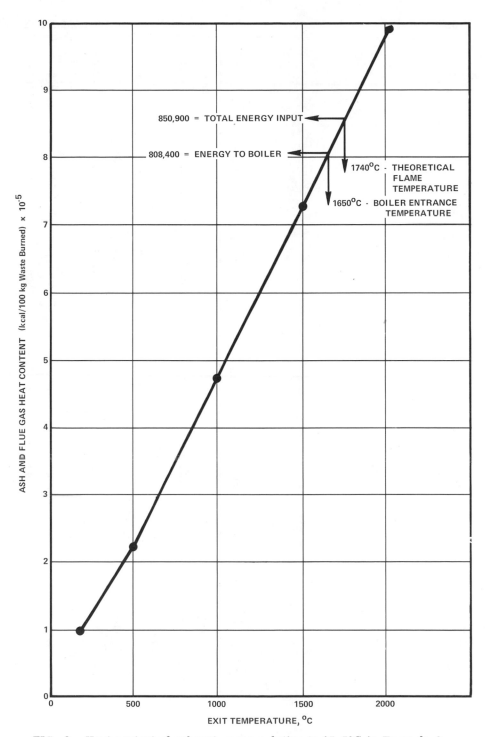

FIG. 3. Heat content of exhaust gases relative to 15.5°C in Example 6.

TABLE 7

Computation of Heat Content of Flue Gases from Combustion of Benzene Waste at 10% Excess Air

		180	500	1000	1500	2000
A	Assumed temp. (°C)	180	500	1000	1500	2000
B	(A) – 15.5°C	164.5	484.5	984.5	1484.5	1984.5
C	$Mc°_{p,\,av}$ N_2[a]	7.00	7.13	7.48	7.76	8.00
D	$Mc°_{p,\,av}$ O_2	7.10	7.50	7.92	8.20	8.40
E	$Mc°_{p,\,av}$ H_2O	8.10	8.52	9.23	9.90	10.50
F	$Mc°_{p,\,av}$ CO_2	9.42	10.75	11.90	12.60	13.08
G	Ash^b	0.2	0.2	0.2	0.2	0.2
H	30.663(B)(C)	35,300	105,900	225,700	353,300	486,800
I	0.741(B)(D)	900	2700	5800	9000	12,400

J	$3.385(B)(E)$	4500	14,000	30,800	49.700	70,500
K	$14.846(B)(F)$	23,000	77,300	173,900	277,700	385,300
L	$9.3[(B)(G) + 85]$[c]	300	900	2600	3600	4500
M	$3.385(10,595)$[d]	35,900	35,900	35,900	35,900	35,900
N	$(H + I + J + K + L + M)$[e]	99,900	224,100	474,700	729,300	995,400
O	kcal/mol of gas	2440	5460	11,560	17,760	24,255

[a]Source: Fig. 1 kcal/kg mol °C.
[b]Specific heat of the ash (kcal/kg °C) for solid or liquid.
[c]The latent heat of fusion of the ash (85 kcal/kg) is added at temperatures greater than 800°C, the assumed ash fusion temperature.
[d]Latent heat of vaporization at 15.5°C of free water in waste and from combustion of hydrogen in waste (kcal/kg mol).
[e]Total heat content of gas stream (kcal).

	kcal	Temperature °C
Energy into system		
Heat of combustion	772,730	15.5
Air preheat	78,170	300
Total	850,900	1740[a]
Heat loss (5%) from combustion chamber	(42,500)	
Energy into boiler	808,400	1650
Heat loss (5%) from boiler	(40,400)	
Heat loss out stack	(99,900)	180
Net energy into steam	668,100	204

[a] The theoretical (adiabatic) flame temperature for this system (the temperature of the products of combustion assuming no heat loss).

For feedwater at 100°C, 15.8 atm changing to saturated steam at 15.8 atm, the enthalpy change is 567.9 kcal/kg so the resulting steaming rate for a burning rate of 1,100 kg/hr is

$$\frac{(668,100)(11)}{567.9} = 12,940 \text{ kg/hr}$$

Such analysis produces a number of other useful measures of the system and design values. These include:

Available results	Utility
Total air rate	Fan and duct sizing
Ash rates	Ash disposal equipment sizing
Combustion chamber	
Temperature	Refractory selection
Gas flow rates	Chamber and duct sizing to obtain residence time, avoid erosion

Available results	Utility
Boiler	
Temperature	Heat transfer, steaming rate
Gas flow rates	Heat transfer, erosion protection
Energy flows	Steaming rate
Stack gas flow rate	Pollution control, duct and fan sizing

Note also that if the level of preheat was a variable, Figure 3 could have been used to determine the preheat needed to obtain (1) a given flame temperature, or (2) a desired steaming rate by the following:

1. Determine energy content of combustion chamber gases corresponding to the flame temperature which: (a) meets criteria which assure complete combustion, or (b) corresponds to an energy differential with that of the stack gas which produces the desired steaming rate.

2. Multiply result by $(1/0.95) = 1.053$ to provide for heat loss in combustion chamber.

3. Subtract the heat of combustion of the waste. The result is the energy deficiency to be satisfied by the preheated air.

4. Divide result by combustion air mole flow rate. The result is the needed molar heat content of the combustion air (kcal/mol).

5. Using Fig. 3, find (by iteration) the air preheat corresponding to the needed heat content derived in item 4:

$$\text{Temperature (T)} = \frac{\text{heat content}}{Mc^{\circ}_{p,av} @T}$$

D. EQUILIBRIUM

It is a recognized property of chemical reactions, readily demonstrable from thermodynamic principles, that no reactions go "to completion." Always, some unreacted materials remain. Further, all other possible chemical reactions involving the primary reactants, the primary products, and all intermediates will take place to some extent.

In most situations of interest, however, many of the thermodynamically possible reactions are neglected, unimportant either because the reaction rates are too slow to produce significant amounts of product, or the quantities of material involved are so small as to be insignificant.

Thermodynamic analysis shows that the extent of the reaction of ideal gases (a realistic assumption at combustion temperatures) is given by the equilibrium constant K_p which is a function only of temperature. The equilibrium constant for the reaction $aA + bB \rightleftharpoons cC + dD$ where the concentrations of reactants and products are expressed as partial pressure is given by

$$K_p = \frac{p_C^c \, p_D^d}{p_A^a \, p_B^b} \tag{11}$$

The dimensions of K_p depend on the algebraic sum of the exponents $(c + d - a - b)$. If the total is zero, K_p is dimensionless. If the total is non-zero, K_p will have the units of pressure raised to the appropriate integer or fractional power.

Figure 4 shows the temperature dependence of reactions of primary interest with K_p based on partial pressures expressed in atmospheres. H_2O is vapor, carbon (solid) is beta-graphite. For combustion calculations, several reactions are worthy of consideration in evaluating heat and material flows, especially the gas phase reactions of H_2, H_2O, CO, CO_2, O_2, with H_2O, CO_2, and O_2. Note that for the carbon reactions, the activity of the solid is assumed to be unity and carbon "concentration" or "partial pressure" does not enter into the mathematical formulation.

The partial pressure of the gases in a mixture at atmospheric pressure can be conveniently equated with the mole fraction (moles of component divided by the total number of moles in the mixture). This equivalence and its use in the equilibrium formulation is strictly true only for ideal gases. Fortunately, at the temperatures and pressures typical of most combustion calculations, this is a valid assumption.

The importance of the reactions described in Fig. 4 relates to both combustion energetics and pollution control. For example, equilibrium considerations under conditions near to stoichiometric may indicate that substantial nitric oxide and carbon monoxide may be generated. Also, the interactions of H_2O, CO, CO_2, C, and O_2 are important in understanding the combustion behavior of char and in arriving at estimates of the unreleased heat of combustion of CO.

Example 7. For the combustion chamber conditions of Example 6, what is the emission rate of NO due to the reaction $1/2\,O_2 + 1/2\,N_2 = NO$? If this result indicates NO emissions are excessive, one approach to control would involve recycle of stack gas at 180°C to dilute and cool the combustion chamber off-gas to 1000°C. What is the flow rate of the recycle stream and the resulting NO emission rate?

It should be noted that the NO emission rate for a real combustor will fall somewhere between the value calculated at combustion chamber effluent temperatures and that for stack temperatures. NO is not formed in the high temperature zones quite to the theoretical equilibrium level due to local temperature or concentration variations and to reaction rate limitations. The NO concentration also does not follow the theoretical equilibrium curve as the gas temperature falls (in traversing the boiler) due to reaction rate limitations. The calculations below, therefore, are a "worst case": theoretical NO formation is the furnace and "frozen equilibrium" through the boiler.

From Example 6, the concentrations of N_2, O_2, and NO are:

Component	Moles	Mole percent (partial pressure)
NO	x	x/40.854
N_2	30.663 − 0.5x	(30.663 − 0.5x)/40.854
O_2	0.741 − 0.5x	(0.741 − 0.5x)/40.854
Total (CO_2, H_2O, O_2, N_2)[a]	40.854	

[a]Note that for this reaction, the total number of moles of gas does not change.

From Fig. 4, at 1650°C, $\log K_p = 1.8$ ($K_p = 63.1$), where $K_p = p_{O_2}^{1/2} p_{O_2}^{1/2}/p_{NO}$, and p_{O_2}, and p_{N_2}, and p_{NO} are the partial pressures of O_2, N_2, and NO, respectively, each expressed in atmospheres. At equilibrium, then,

$$(0.741 - 0.5x)^{1/2}(30.663 - 0.5x)^{1/2} = 63.1x$$

Solving this equation gives x = 0.074 mol of NO at equilibrium or 0.180 mol % or 1800 parts per million (ppm) by volume. Because the heat of formation of NO is −21,600 kcal/mol of NO formed, this reaction will absorb 1590 kcal or 0.21% of the heat of combustion calculated above.

After review of state and local regulations, it is judged that NO emissions (regulated as "NO_x" and calculated as NO_2 since some oxidation to NO_2 will occur) could become excessive and the recycle option should be

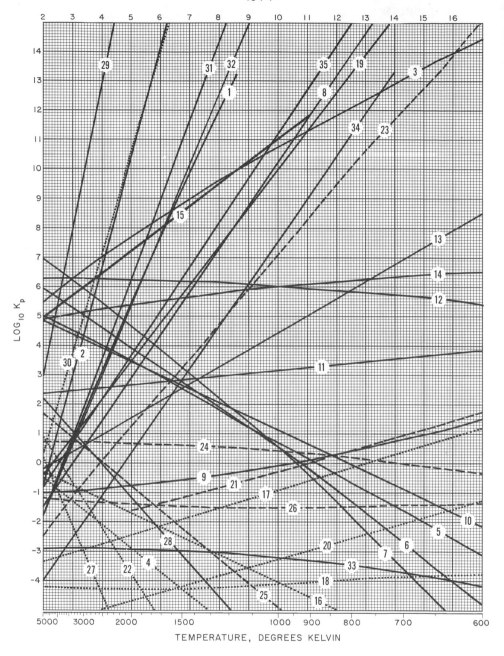

FIG. 4. Equilibrium constants for reactions common in combustion systems.

38

Reaction Equilibria Shown in Figure 4.

A. Carbon Reactions

1. $\frac{1}{2}C_2(g) \longrightarrow C(s)$

8. $CO + \frac{1}{2}O_2 \longrightarrow CO_2$

2. $C(g) \longrightarrow C(s)$

9. $CO + H_2O \longrightarrow CO_2 + H_2$

3. $C + \frac{1}{2}O_2 \longrightarrow CO$

10. $CH_4 \longrightarrow C + 2H_2$

4. $C + \frac{1}{2}N_2 \longrightarrow \frac{1}{2}C_2N_2$

11. $\frac{1}{2}C_2H_4 \longrightarrow C + H_2$

5. $C + 2H_2O \longrightarrow CO_2 + 2H_2$

12. $HCHO \longrightarrow CO + H_2$

6. $C + H_2O \longrightarrow CO + H_2$

13. $\frac{1}{2}C_2H_2 \longrightarrow C + \frac{1}{2}H_2$

7. $C + CO_2 \longrightarrow 2CO$

14. $\frac{1}{3}C_3O_2 + \frac{1}{3}H_2O \longrightarrow CO + \frac{1}{3}H_2$

C. Nitrogen and Oxygen Reactions

15. $O_3 \longrightarrow \frac{3}{2}O_2$

18. $NO + \frac{1}{2}N_2 \longrightarrow N_2O$

16. $\frac{1}{2}N_2 + \frac{1}{2}O_2 \longrightarrow NO$

19. $H_2 + \frac{1}{2}O_2 \longrightarrow H_2O$

17. $NO + \frac{1}{2}O_2 \longrightarrow NO_2$

20. $\frac{1}{2}N_2 + \frac{3}{2}H_2 \longrightarrow NH_3$

C. Sulfur Reactions

21. $\frac{1}{2}S_2(g) \longrightarrow S(\ell)$

24. $\frac{1}{3}SO_2 + \frac{2}{3}H_2S \longrightarrow \frac{1}{2}S_2(g) + \frac{2}{3}H_2O$

22. $SO_2 \longrightarrow SO + \frac{1}{2}O_2$

25. $H_2S \longrightarrow HS + \frac{1}{2}H_2$

23. $SO_2 + 3H_2 \longrightarrow H_2S + 2H_2O$

26. $CO + H_2S \longrightarrow COS + H_2$

D. Radical Reactions

27. $C + \frac{1}{2}N_2 \longrightarrow CN$

31. $2O \longrightarrow O_2$

28. $CH_4 \longrightarrow CH_3 + \frac{1}{2}H_2$

32. $2H \longrightarrow H_2$

29. $2N \longrightarrow N_2$

33. $\frac{1}{2}H_2 + O_2 \longrightarrow HO_2$

30. $N + O \longrightarrow NO$

34. $OH + O \longrightarrow HO_2$

35. $OH + \frac{1}{2}H_2 \longrightarrow H_2O$

considered. Also, since we are trying to minimize combustion temperature, cold (15.5°C) air will be used.

Basis: 100 kg of waste

For ease in computation, assume the two-step system in Fig. 5. Since no energy will enter the system with the air, the heat in is 772,730 kcal, of which 734,100 kcal will be found in the combustion gas. With 5% heat loss, and from Fig. 3, the temperature will be 1510°C. This is also the temperature for Example 6 with no air preheat.

Drawing a balance around the mixing chamber, let n = number of moles of recycled gas.

The heat content of the net products of combustion at 1000°C is found from Fig. 3 to be 474,700 kcal. Since the recycle gas has the same chemical composition as the products of combustion, the sensible heat content of the recycle gas at 1000°C is in the same ratio to the sensible heat content of the products of combustion at 1000°C as the number of moles of recycle is to the number of moles of net combustion products. At 1000°C, then, the heat content of the recycle gas is $474,700 (n/40.854) = 11,619n$. Since the slag heat does not appear in the recycle gas, however, a more accurate value can be taken from Table 7: $11,560n$.

To satisfy an energy balance around the combustion chamber, the energy in the fuel plus that in the air (zero in this case) plus that in the recycle at 180°C (see Table 7) must equal that of the mixed gas stream at 1000°C.

$$734,100 + 0 + 2440n = 474,700 + 11,560n$$

$$n = 28.44 \text{ mol}$$

For the actual case, at eleven times the flow used as the basis for the example, 312.84 mol/hr would be recycled.

Under the new and lower exhaust temperature, the NO generation rate may be recalculated. Noting that the recycle gas is of the same composition as the original furnace gas, the new equilibrium calculation differs from that given above only in that K_p is that for 1000°C; log $K_p = 3.05$ at 1000°C, $K_p = 1122$. Solving as above, only 0.00424 mol of NO are now formed at equilibrium, or 0.010 mol %, or 100 ppm.

E. COMBUSTION KINETICS

Combustion processes, as for other chemical reactions, proceed at a finite rate dependent upon temperature, the concentrations of the reacting species, and in some cases, the static pressure. The description of the relationship

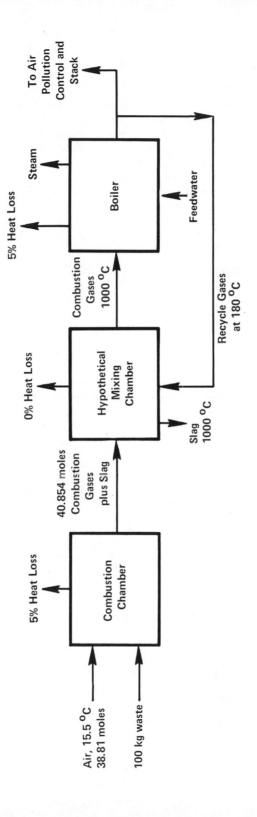

FIG. 5

between reaction rate and system parameters is the subject matter of combustion kinetics.

It should be recognized that in most practical combustors reaction rate does not or should not control the burning rate. In the more typical case, temperatures are maintained high enough that reactions proceed rapidly in comparison to the mean residence time and other, slower processes are rate-limiting; usually the mixing rate of fuel and oxidant (both eddy and molecular). It is useful, however, for the incinerator designer to have a basic understanding of the principles involved and to study the kinetics of combustion for two important species: carbon monoxide and soot (carbon).

1. Introduction to Kinetics

a. Overall Kinetics

For the generalized, gas phase chemical reaction

$$bB + cC \rightleftharpoons dD + eE$$

The rate of the forward reaction (to the right) has been found to often follow the form

$$r = k[B]^b[C]^c \tag{12}$$

where the value in the brackets (for ideal gases) is the partial pressure of the component.

Similarly, the back reaction rate (to the left)

$$r' = k'[D]^d[E]^e \tag{13}$$

At equilibrium, by definition, the forward reaction and the backward reactions proceed at the same rate:

$$r = r'$$

$$k[B]^b[C]^c = k'[D]^d[E]^e$$

$$\frac{k}{k'} = \left(\frac{[D]^d[E]^e}{[B]^b[C]^c}\right)_{equilibrium}$$

$$= K_p \text{ (the equilibrium constant)} \tag{14}$$

Van't Hoff, from thermodynamic reasoning, examined the variation of the equilibrium constant K_p with temperature T and concluded

$$\frac{d \ln K_p}{dT} = \frac{\Delta H}{RT^2} \tag{15}$$

where ΔH is the heat of reaction (kcal/mol) and R is the gas constant.

Combining the two previous expressions:

$$\frac{d \ln k}{dT} - \frac{d \ln k'}{dT} = \frac{\Delta H}{RT^2} \tag{16}$$

The right side of this equation could be divided into two parts, provided the overall heat of reaction is broken into an energy change for each direction:

$$\Delta H = E - E' \tag{17}$$

or

$$\frac{d \ln k}{dt} - \frac{d \ln k'}{dt} = \frac{E}{RT^2} - \frac{E'}{RT^2} \tag{18}$$

The two separate expressions, one for the forward and one for the backward reaction, having a difference in agreement with the just-stated equilibrium requirement are

$$\frac{d \ln k}{dt} = \frac{E}{RT^2} \tag{19}$$

$$\frac{d \ln k'}{dt} = \frac{E'}{RT^2} \tag{20}$$

Integration yields the Arrhenius equation, Eq. (21):

$$k = Ae^{-E/RT} \tag{21}$$

The quantity E was interpreted by Arrhenius as an excess over the average energy which the reactants must possess in order to allow the reaction to occur. He termed this the activation energy (actually an enthalpy) and related its significance to the energy required to form an activated intermediate between reactants and products. One might picture such an intermediate as the energy-rich assembly of two colliding molecules just before bonding electrons had shifted to form some product. The only constraint on E and E' is that their difference be equal to the net heat of reaction—their individual magnitudes depend on the particular structures of the molecules involved.

The term A (the "pre-exponential factor") is interpreted as a combination of molecular collision frequency parameters, steric (orientation) factors and other influences.

The energy changes as a reaction proceeds are illustrated in Fig. 6 for both exothermic (heat releasing) or endothermic (heat absorbing) reactions.

The significant trends which one can draw from an examination of the Arrhenius expression are:

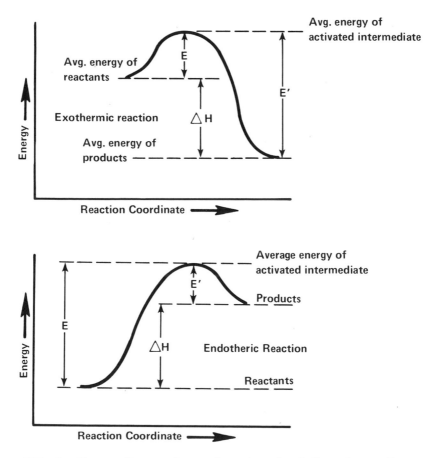

FIG. 6. Energy diagram for exothermic and endothermic reactions.

Rate always increases as temperature increases.

For large ΔE (more typical of exothermic combustion reactions), the increase in rate can be very dramatic over a relatively narrow temperature range. This gives rise to the useful concept of an "ignition temperature."

The Arrhenius concept, developed in 1889 [10], has been used successfully to correlate a wide variety of kinetic data (i.e., by plotting ln k vs. 1/T to give a straight line). Indeed, all failures to follow such a form have been successfully traced to previously unknown side reactions or catalytic effects.

b. Mechanism

The formulation of reaction rate expressions from the simple stoichiometric relationships drawn from the equation for the reaction may be in error. In such cases, the error may often be eliminated when the full sequence of reactions leading from reactants to products is more fully understood. Exploring the true route of a reaction is the study of mechanism. Such studies, particularly in combustion systems, often indicate the vital role of highly activated species, fragments of molecules, called free radicals. These free radicals, existing individually for only fractions of a second, are nonetheless important intermediates in the initiation and propagation of combustion reactions.

Example 8. As an example of the complexity of reaction mechanisms, consider the well-studied gas phase reaction of hydrogen and bromine. The stoichiometric reaction is:

$$H_2 + Br_2 \longrightarrow 2\,HBr$$

Experimental data indicate that the rate of formation of HBr is given by:

$$\frac{dC_{HBr}}{dt} = \frac{kC_{H_2}C_{Br_2}^{1/2}}{1 + k'(C_{HBr}/C_{Br_2})}$$

where C_x indicates the concentration of species x. This expression can be explained by assuming that the reaction proceeds by the following series of steps (note that observation of the steps is impossible and thus these steps are inferred based on their success in explaining the rate data):

Step 1 $1/2\,Br_2 \longrightarrow Br\cdot$

Step 2 $H_2 + Br\cdot \longrightarrow HBr + H\cdot$

Step 3 $H\cdot + Br_2 \longrightarrow HBr + Br\cdot$

Step 4 $H\cdot + HBr \longrightarrow H_2 + Br\cdot$

Step 5 $2Br\cdot \longrightarrow Br_2$

Step 1 is the initiating step, generating the reactive bromine free radical
($Br\cdot$). Steps 2 and 3, occurring over and over in a chain reaction, are
propagating steps, producing product and reseeding the reaction mass with
radicals. Since the concentration of $H\cdot$ in Step 3 is very small, the 0.5
power dependency on bromine concentration and the unity power on hydrogen
concentration in the numerator of the overall rate equation appears reasonable.

Steps 4 and 5 are reactions which reverse or terminate the desired
reaction processes, respectively. As HBr concentration builds, Step 4 becomes important, as indicated by the term in the denominator. If higher
initial bromine concentrations are used, the importance of Step 5 is depressed, resulting in the compensatory bromine concentration term in the
denominator.

An interesting example of the importance of free radical reactions in
combustion processes is given by the differing behavior of "heptane" (C_7H_{16})
and "octane" (C_8H_{18}) in automobile engines. Heptane, as used as a standard
test fuel, is a seven-carbon, straight chain hydrocarbon. After losing a
hydrogen atom (becoming the n-heptyl radical) straight chain hydrocarbons
are relatively active radicals and readily enter into secondary combustion
reactions. Thus, the combustion wave expands rapidly throughout the
cylinders: so rapidly that a serious "knock" is experienced. n-Heptane
carries an octane number (indicating relative knock potential) of zero.

Octane, as used as a standard test fuel, is a highly branched molecule,
more properly named 2,2,4-trimethylpentane. It is a characteristic of
branched chains (and the aromatic ring compounds also found in unleaded
but high octane gasolines) that free radicals are "stabilized;" thus losing a
measure of reactivity, slowing combustion, and reducing knock. 2,2,4-Tri-
methylpentane carries an octane number of 100.

Tetraethyl lead acts as a reaction inhibitor, reducing the radical concentration although at the price of (1) increasing pollution from lead emissions, and (2) reducing the activity (poisoning) of some catalytic converter
catalysts.

Inhibition or radical stabilizing activity is also shown by chlorine-
containing organic compounds, a fact which accounts for problems in obtaining complete combustion of such compounds in combustion systems opti-
mized for nonhalogenated hydrocarbons.

Seldom does the analyst of waste incineration systems have the opportunity or the need to study the fine detail of his process and thus to explore

such kinetic relationships. However, having a basic understanding of the fundamental processes involved can be useful in interpreting problems even if the system under study is too complex to allow an a priori evaluation of reaction rate problems.

2. Kinetics of Carbon Monoxide Oxidation

Carbon monoxide is an important air pollutant, a poisonous gas in high concentrations, and can be the unwanted repository of considerable combustion energy in inefficient combustors. In many instances where hydrocarbons are burned, the oxidation reactions proceed rapidly to the point where CO is formed and then slow greatly until CO burnout is achieved. Indeed, the kinetics of methane oxidation almost duplicate the CO kinetic expressions.

CO is produced by the incomplete combustion of pyrolysis products of solid or liquid wastes, from the char in a refuse bed, or as an intermediate combustion product. The oxidation kinetics to carbon dioxide have been studied by several investigators and, although there are differences in the rate constant reported, a reasonable estimate can be made of the times required to complete the combustion. The rate expression by Hottel et al. [11] will be used to calculate the kinetics of oxidation in a combustion chamber for the condition where the CO and oxygen (air) are intimately mixed. The kinetic expression for the rate of change of CO mole fraction (f_{CO}) with time is given by

$$\frac{-df_{CO}}{dt} = 12 \times 10^{10} \exp\left(\frac{-16,000}{RT}\right) f_{O_2}^{0.3} f_{CO} f_{H_2O}^{0.5} \left(\frac{P}{R'T}\right)^{1.8}$$

where f_{CO}, f_{O_2}, and f_{H_2O} are the mole fractions of CO, O_2, and water vapor, respectively, T is the absolute temperature (K), P is the absolute pressure (atm), t the time in seconds, R is the gas constant (1.986 kcal/kg mol K) and R' is also the gas constant, but in alternate units (82.06 atm cm^3/g mol K).

In reviewing the reaction rate expression, it is instructive to note the dependence upon the mole fraction of water vapor. This results from the participation of hydrogen and hydroxyl (OH) free radicals in the reaction.

To the extent that the combustion chamber may be considered as an isothermal reactor, the decline in CO concentration over an interval t (seconds) from an initial mole fraction $(f_{CO})_i$ to a final mole fraction $(f_{CO})_f$ is given by

$$\frac{(f_{CO})_f}{(f_{CO})_i} = \exp(-Kt) \tag{23a}$$

where

$$K = 12 \times 10^{10} \exp\left(\frac{-16,000}{RT}\right) f_{O_2}^{0.3} f_{H_2O}^{0.5} \left(\frac{P}{R'T}\right)^{0.8} \tag{23b}$$

Other workers have concluded that Eq. (22) gives unrealistically high rates at low temperatures (e.g., in afterburners operating below 1000°C). A more conservative result is given by Morgan [11a]:

$$\frac{-df_{CO}}{dt} = 1.8 \times 10^{13} f_{CO} f_{O_2}^{0.5} f_{H_2O}^{0.5} \left(\frac{P}{R'T}\right)^{0.5} \exp\left(\frac{-25,000}{RT}\right)$$

Example 9. 470 m^3/min of a waste gas stream from a carbon char reactivation furnace is produced at 300°C. Its composition is:

Component	Mole percent
N_2	78.17
CO_2	18.71
CO	2.08
O_2	1.04
Total	100.0

The waste gas will be mixed with the flue gas from an atmospheric natural gas (methane) combustor operating at 50% excess air in a "fume incinerator" or afterburner. If CO combustion is assumed not to begin until the two gas streams are intimately mixed, plot the required residence time to reduce CO concentration to 10 ppm (neglecting dilution effects) against the mixed-gas temperature. Note that residence time is directly related to the incinerator volume (capital cost) and mixed-gas temperature to methane consumption rate (operating cost). For ease in computation, assume the heat capacity of the two gas streams are constant, identical, and equal to about 8 kcal/kg mol °C. Neglect heat release from CO combustion. Let the datum for sensible heat content be 15.5°C. The system diagram is shown below:

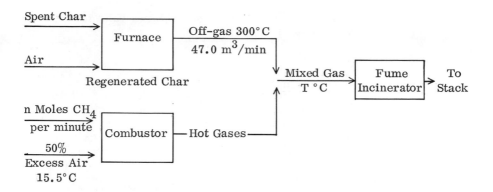

For 1 mol of CH_4, at 50% excess air, the hot combustor gas produced has the composition:

Component	Moles
N_2	11.29
CO_2	1.00
O_2	1.00
H_2O	2.00
Total	15.29

The molal flow rate of gases from the furnace is

$$470\left(\frac{273}{300 + 273}\right)\left(\frac{1}{22.4}\right)= 10 \text{ mol/min}$$

Basis: 1 min of operation with a methane firing rate of n moles per minute.

  Mixed Gas Temperature

1. Total heat content

    a. Sensible heat in furnace gas = $10 \ \overline{Mc_p} (300 - 15.5) = 22,760$ kcal

b. Heat of combustion of methane (lower heating value from Table 6) = $(191,910)n$ kcal

2. Mixed-gas temperature $(T\ ^\circ C)$

$$T = \frac{\text{Total heat content}}{(\overline{Mc}_p)\ (\text{no. moles})} + 15.5 = \frac{22,760 + 191,910n}{8(10 + 15.29n)} + 15.5$$

Mixed Gas Composition

Component	From furnace	From methane	Total
		Moles	
N_2	7.82	11.29n	7.82 + 11.29n
CO_2	1.87	n	1.87 + n
O_2	0.10	n	0.1 + n
H_2O	0.0	2.0 n	2.0n
CO	0.21	0	0.21
	10.0	15.29n	10 + 15.29n

Oxygen mole fraction $= f_{O_2}^* = \dfrac{0.1 + n}{10 + 15.29n}$

Water mole fraction $= f_{H_2O} = \dfrac{2n}{10 + 15.29n}$

CO Reaction Requirement

$(f_{CO})_i$ = Initial CO mole fraction $= 2.08 \times 10^{-2}$

$(f_{CO})_f$ = Final CO mole fraction $= 10^{-5}$

$$\frac{(f_{CO})_f}{(f_{CO})_i} = \frac{10^{-5}}{2.08 \times 10^{-2}} = 4.81 \times 10^{-4} = \exp(-Kt)$$

$Kt = -\ln(4.81 \times 10^{-4})$

---

*Neglect, for simplicity, the change in O_2 concentration due to CO combustion and changes in the total moles in the system.

$$Kt = 7.64$$

$$t = \frac{7.64}{K} \text{ sec}$$

Reaction Time and Methane Usage. From the relationships developed above, the reaction rate parameter K and the required residence time t can be calculated as a function of n, the methane consumption rate. Then, for a given value of n, and the associated total mole flow rate and absolute temperature (T'), the volume flow rate of the mixed gas $\dot{V}$ can be calculated:

$$\dot{V} = (\text{mol/min}) \frac{22.4}{6} \frac{T'}{273} \text{ m}^3/\text{sec}$$

The chamber volume V is then given by:

$$V = \dot{V} t \text{ m}^3$$

The calculations are shown in Table 8 and the results are plotted in Figure 7.

TABLE 8

Calculations for Example 9

n (Moles CH_4/min)	0.2	0.4	0.6	0.8	1.0
Total moles	13.06	16.12	19.17	22.23	25.29
Temperature (°C)	600	787	899	991	1076
T' (K)	874	1060	1172	1264	1350
f_{O_2}	0.023	0.031	0.037	0.040	0.043
f_{H_2O}	0.031	0.050	0.063	0.072	0.079
K (sec^{-1})	88	529	1187	2038	3095
t (sec)	0.087	0.014	0.006	0.004	0.002
$\dot{V}$ (m^3/sec)	15.6	23.4	30.7	38.4	46.7
V (m^3)	1.360[a]	0.337	0.198	0.144	0.115

[a] Note that if the oxygen mole fraction corresponding to complete combustion of CO had been used ($f_{O_2} = 0.09$) instead of $f_{O_2} = 0.023$, the volume would have increased to 1.95 m^3

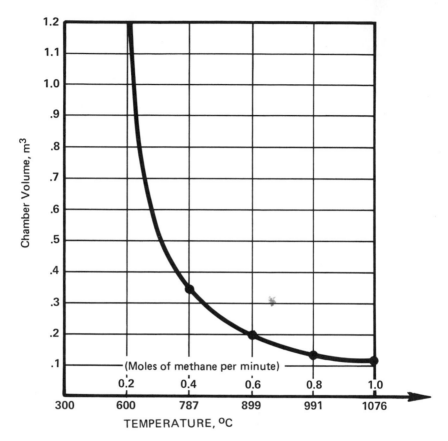

FIG. 7. Capital versus operating cost trade-off analysis for CO after-burner (Example 9).

Several points become clear upon review of Table 8 and Figure 7:

The strong influence of temperature, typical of kinetically controlled reactions

It would appear that around 650°C, the combustion rate of CO increases sharply. This could be considered an "ignition temperature" for the low oxygen concentration mixture existing in the afterburner. At normal atmospheric oxygen concentrations, the CO ignition temperature [12] is approximately 600°C (note that for $f_{O_2} = 0.2$, the chamber volume is only about $0.7 \ m^3$ at 600°C and a V-n curve such as shown in Fig. 7 would just be breaking sharply upward).

3. Kinetics of Soot Oxidation

In burning carbon bearing wastes, conditions of high temperature and low oxygen concentration can lead to the formation of soot. The mechanisms responsible for soot formation include preferential oxidation of the hydrogen and thermally induced dehydrogenation. Whatever the cause, soot formation is a problem to the incineration system:

Dark, high optical density flue gas emissions which, though less than the applicable mass emission limits, may violate opacity restrictions.

It is a sign of poor combustion conditions and, most likely, is accompanied by high carbon monoxide emissions.

If metal surfaces (e.g., the fire-side of boiler tubes) are exposed to atmospheres which swing from oxidizing to reducing conditions (the regions where soot is formed are often reducing), rapid metal wastage occurs.

If an electrostatic precipitator is used for air pollution control, the presence of carbon lowers the resistivity of the dust and may lower collection efficiency. In fabric filter control devices, capture of the slow-burning char may greatly decrease bag life.

Unfortunately, once soot is formed, its slow burnout rate makes subsequent control efforts difficult.

The control of soot burnout can best be understood by examining the kinetics of combustion of carbonaceous particles. For spherical particles, review of the considerable research on this topic [13] suggests the following:

$$q = \frac{p_{O_2}}{1/K_s + 1/K_d} \tag{24}$$

where q is the rate of carbon consumption (g/cm^2 sec^{-1}), p_{O_2} is the partial pressure of oxygen (atm), K_s is the kinetic rate constant for the consumption reaction, and K_d is the diffusional rate constant. Both kinetic and diffusional resistances to reaction are thus seen to influence burnout rate.

For (small) particles of diameter d (centimeters) typical of soot, the diffusional rate constant at temperature T (K) is approximately given by:

$$K_d = \frac{4.35 \times 10^{-6} T^{0.75}}{d} \tag{25}$$

The kinetic rate constant is given by:

$$K_s = 0.13 \exp\left[\left(\frac{-35,700}{R}\right)\left(\frac{1}{T} - \frac{1}{1600}\right)\right]$$ (26)

where R is the gas constant (1.986 cal/g mol K).

For a particle of initial diameter d_0 and an assumed specific gravity of 2, the time (t_b) in seconds to completely burn out the soot particle is given by:

$$t_b = \frac{1}{P_{O_2}}\left[\frac{d_0}{0.13 \exp\left[\left(\frac{-35,700}{R}\right)\left(\frac{1}{T} - \frac{1}{1600}\right)\right]} + \frac{d_0^2}{8.67 \times 10^{-6}T^{0.75}}\right]$$ (27)

Example 10. The combustion of benzene and other unsaturated ring compounds is known to often result in the formation of soot. If soot is formed at 1000°C in Example 7, what are the burnout times for 2-, 20-, and 200-μm (1 μm = 10^{-4} cm) soot particles?

From Example 7, the oxygen partial pressure, corrected for NO formation, is [0.741 - 0.5(0.074)]/40,854 = 0.0172 atm. For this condition, at a temperature of 1273K, the burnout equation reduces to:

$$t_b = 58\left(\frac{d_0}{2.33} + \frac{d_0^2}{0.00185}\right)$$

$$\text{Kinetic term} \qquad \text{Diffusional term}$$

and:

Initial diameter (μm)	Burn-out time (sec)
2	0.006 (kinetic control)
20	0.175 (mixed control)
200	13.052 (diffusion control)

It can be seen that small soot particles (slightly above 20 μm in diameter) could be expected to burn out in the 0.5 to 1 sec residence time in the boiler hot zone (before a significant temperature drop occurs as the gases pass over the cool boiler tubes). Large particles or agglomerates could be a problem, since the diffusional resistance to carbon consumption is high and is the predominant influence in extending the burnout time.

Chapter 3

SELECTED TOPICS ON COMBUSTION PROCESSES

Combustion processes involve mass and energy transport and chemical reaction. Depending on the relative pace of these necessary steps, the process
changes in its characteristics. Such changes can affect the degree of success of the combustion system in performing according to expectations.

It is the goal of design engineering to be able to apply an understanding
of the fundamental physical and chemical processes which proceed in systems of interest in order to quantitatively predict performance and, conversely, to translate performance goals into system hardware designs. In
the design of combustion systems, such a first-principles approach is both
difficult and time consuming. Yet exploration of the processes involved
gives valuable insight into the appropriateness of designs and into the interpretation of problems.

To this end, complete combustion processes for gaseous, liquid, and
solid fuels are described. Also, intentionally incomplete combustion (or
pyrolysis) is described to aid in understanding several waste processing
concepts undergoing development and implementation in the mid-1970s
which are based on this concept.

A. GASEOUS COMBUSTION

Although most wastes requiring incineration are not gaseous, it should be
recognized that most combustion is completed in gas phase reactions. Many
liquids vaporize in the hot furnace environment, and it is the gasified liquid
which actually engages in combustion. Heavier liquids and solids are pyrolyzed by intense heating, releasing volatile fragments which burn in the gas

phase. Even carbon char (appearing as the waste of interest or as a product of the pyrolysis process) is usually gasified (by oxygen or water vapor) to carbon monoxide which burns in the gas phase. In reviewing gas-phase combustion, systems may be divided into those where fuel and stoichiometric oxidant enter the combustion environment separately and where the combustion rate is almost always mixing limited, and those where combustion is initiated in homogeneous fuel-oxidant mixtures, with the flame reactions propagating through the system in a substantially continuous (although, perhaps, smudged by turbulence) flame front. The latter case is found primarily only in premixed gas burners. The former is by far the most common.

The mechanism by which oxidant and fuel are brought together provides a second classification of flame type. When molecular diffusion predominates, the flame is known as a laminar diffusion flame. When eddy diffusion predominates, the flame is known as a turbulent diffusion flame. As with most flow processes, the Reynolds number of the stream, a measure of the ratio of momentum to viscous forces, is a useful indicator of the regimes where one or the other flow conditions exist.

1. The Premixed (Bunsen) Laminar Flame

Visual examination of simple, premixed conical hydrocarbon-air flames indicates three regions: the preluminous, luminous, and postluminous. In the preluminous zone, the cold feed gas undergoes preheating and some "seeding" with reactive species due to thermal and mass diffusion against the direction of convective flow. At some point the combination of temperature and mixture reactivity reaches a level where rapid heat release begins. Activated radicals (e.g., OH·, C=C, and others) appear in high concentration and account for much of the visible light released in the luminous zone. Gouy suggested in 1881 [14] that the conical shape of the luminous zone could be explained by assuming that the flame propagates normal to itself at a constant rate, known as the burning velocity or flame speed. This simple concept has withstood the testing of many investigators and appears to be able to explain quantitatively flame front geometry. For most hydrocarbon fuels, the flame speed falls in the range from 0.36 to 0.58 m/sec at normal atmospheric temperatures and pressures.

It should be recognized that the combustion reaction is seldom complete within the luminous zone. Particularly for the combustion of complex fuels, heat release from CO and H_2 oxidation extends into the postluminous zone (an afterburning region). In the postluminous zone, the temperature falls due to radiation heat loss and to the assimilation of cool ambient fluid into the flame flow, and combustion is completed.

Figure 8 illustrates the heat release rate for a stoichiometric propane-air flame at 0.25 atm [15]. The low pressure increases the thickness of the

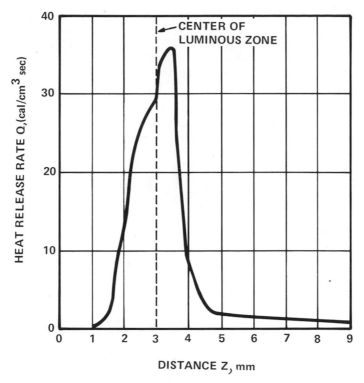

FIG. 8. Heat release rate in stoichiometric propane–air flame at 0.25 atm. Curve calculated from the temperature profile using estimated thermal conductivities.

luminous zone. A similar thickening is observed as the degree of premix (air-to-fuel ratio) is decreased.

2. The Diffusion Flame

In a diffusion flame, fuel combustion reactions must await the arrival of oxidant. Preparatory steps such as pyrolysis, evaporation, or thermal cracking, however, may proceed under the influence of radiative heat transfer from the (typically) hot surroundings, the heating due to turbulent mixing of hot burned gases into the fuel stream, and conduction from the flame regions.

For most burner systems, the fuel is introduced into the combustion chamber in a jet, with or without swirl. Since combustion is mixing limited,

the flow characteristics (especially entrainment) of the jet are prime deter-
minants of combustion rate, flame length, and flame shape. Since jet behavior is very important to diffusion flame and combustor fluid mechanics,
it is discussed in detail below. One important flame quantity, the flame
length, will be mentioned here for the more common case, the turbulent
diffusion flame.

Based on a simplified model for jet mixing, Hottel [16] proposed and
tested the following relationship defining the flame length L, corrected for
the distance X^+ from the nozzle face to the point where turbulence is initiated:

$$\frac{L - X^+}{d} = \frac{5.3}{f_s} \left\{ \left[f_s + (1 - f_s) \left(\frac{M_a}{M_0} \right) \right] \frac{T_F}{\theta_m T_0} \right\}^{1/2} \tag{28}$$

where

d = the nozzle diameter

f_s = the mole fraction of nozzle gas in a stoichiometric gas-air
mixture

M_a, M_0 = the molecular weight of air and nozzle gas, respectively

T_F = the adiabatic flame temperature (Kelvin) of a stoichiometric
mixture

T_0 = the nozzle gas temperature (Kelvin)

θ_m = moles of reactants per mole of product in a stoichiometric
mixture

Estimation of the distance X^+ to the break point is necessary if total flame
length is to be determined. Generally, X^+ is small compared to $(L - s)$
(except for CO flames for which $X^+/(L - X^+)$ is as high as 0.4). On the average, X^+ is about 0.05L, is less than 0.09L in the turbulent range, and
decreases with increasing nozzle Reynolds number.

This correlation is useful in that it allows anticipation of flame jet impingement problems, and allows estimates of the effect of fuel changes
(changes in f_s, M_0, T_F, and θ_m) and of preheat (changes in T_0) on flame
length. Its accuracy for a variety of gaseous fuels ranged within 10 to 20%
of the experimental values. It should be recognized that devices which increase the rate of entrainment of oxidant will act to shorten the flame. Thus,
swirl in particular will significantly shorten flame length.

Example 11. Natural gas at 15°C (methane, with a flame temperature of
2170 K) is to be burned in air from a port 4 cm in diameter at a velocity of
3 m/sec. The furnace is 8 m wide. If flame impingement is foreseen, the
plant engineer suggests cutting back on the fuel flow by up to 20%. Will this
work? What will?

First, the jet should be evaluated to determine if it is, indeed, turbulent.

$$\text{Reynolds no.} = \frac{d_0 u_0 \rho_0}{\mu_0} \qquad (29)$$

where

d_0 = nozzle diameter = 4 cm

u_0 = discharge velocity = 3 m/sec = 300 cm/sec

ρ_0 = density of methane at 15°C = 6.77×10^{-4} g/cm^3

μ_0 = viscosity of methane at 15°C = 2.08×10^{-4} g/cm sec

$$\text{Reynolds no.} = N_{Re} = \frac{(4)(300)(6.77 \times 10^{-4})}{2.08 \times 10^{-4}}$$

$$= 3900 \text{ (dimensionless)}$$

Since the Reynolds number exceeds that required for turbulent flow (2100), the flame length relationship should hold.

Since 1 mol of methane burns stoichiometrically with 2 mol of oxygen, and $2(79/21) = 7.52$ mol of nitrogen, $f_s = 1/(1 + 2 + 7.52) = 0.095$.

Methane burns according to the reaction

$$CH_4 + 2O_2 \longrightarrow CO_2 + 2H_2O$$

$\qquad$ (3 mol) $\qquad\qquad\qquad$ (3 mol)

so

$$\theta_m = 3/3 = 1.$$

Assuming that X^+ is much less than L

$$\frac{L}{d} = \frac{5.3}{0.095} \left\{ \left[0.095 + (1 - 0.095)\left(\frac{29}{16}\right) \right] \frac{2170}{(1)(288)} \right\}^{1/2}$$

$$= 202$$

and

$$L = 8.07 \text{ m}$$

Will dropping the firing rate help? No. The lower inlet velocity will lower the air entrainment rate in proportion, and the flame length will remain essentially constant. What could be done?

1. Use a 3-cm nozzle diameter. This will cut the flame length to 3/4 of the base case with an increase in gas compression costs. Alternatively, use two or more nozzles.

2. Preheat the fuel. If the methane is introduced at 250°C (523 K), the flame length will decrease to 6 m.

3. Partially premix air with the gas (changing f_S and M_0), being careful to avoid flashback problems.

4. Explore the possibility of tilting the burner.

5. Explore using swirling type burners.

Use of the equation presented should cause no problems in "clean" situations. If, however, the burner discharges into hot furnace gases where the oxygen concentration is depleted, or operates in a bank of fuel nozzles where the zone of influence of fuel jets overlaps, a lengthening of the flame can be anticipated.

B. LIQUID COMBUSTION

The combustion of liquids beings with the application of heat flux from the flame, from hot refractory walls, from recirculating burned gases, or from a pilot burner, to vaporize the fuel. The fuel vapors then diffuse from the liquid surface, mixing with oxygen in the air and increasing in temperature. At some point, the rate of oxidation reactions and associated heat release is high enough that full ignition and combustion ensues.

In most situations it is desirable to subdivide the oil into small droplets in order to enhance vaporization rate (increasing the combustion rate per unit volume), to minimize smoking, and to ensure complete combustion. In some simple "pot type" heaters, in pit incinerators, and occasionally in highly unusual waste disposal situations, the sludges or liquids are burned in a pool.

1. Pool Burning

Fires above horizontal liquid pools have been studied both experimentally and analytically to clarify the relationship between burning rate and the properties and characteristics of the system.

In small pools (up to 3 cm in diameter), burning is laminar and, unexplainedly, the burning rate per unit area $\dot{m}''$ decreases in approximate proportion as diameter increases. In a 3- to 5-cm diameter pool the flame is still laminar but regular oscillations occur.

For diameters greater than 7 cm and less than 20 to 30 cm, a transition to turbulent burning takes place. Convection is the primary source of feedback energy influencing $\dot{m}''$ in this pool size range, and the overall $\dot{m}''$ is essentially independent of pool diameter. For flames which yield little radiation (e.g., methanol), the constant $\dot{m}''$ range extends to diameters over 100 cm.

A constancy of $\dot{m}''$ (average) for a range of pool diameters does not imply a constant $\dot{m}''$ with radius. Akita and Yumoto [17] report a 44% maximum radial mass flux variation for a 14.4-cm diameter methanol fire, and Blinov and Khudyakov [18] report an 18%, 18%, and 27% maximum radial variation for benzene, gasoline, and tractor kerosene, respectively, in a 30-cm tank. These variations, however, may be considered small.

Studies by DeRis and Orloff on pool burning rates in the convectively controlled region [19] identified a relationship between burning rate and characteristics of the fuel which appears to satisfactorily correlate several sets of data for values of B (a dimensionless mass transfer driving force) greater than unity. Below B = 1, the burning rate $\dot{m}''$ is lower than predicted, apparently due to the onset of unexplained extinction processes.

DeRis and Orloff's correlation is

$$\dot{m}'' = 0.15 \left(\frac{\lambda_1}{C_{p1}}\right) \frac{g(\rho_a - \rho_3)}{\mu_2 \alpha_2}^{1/3} B\left[\frac{\ln(1 + B)}{B}\right]^{2/3} \quad \text{kg/sec m}^2 \tag{30a}$$

and

$$B = \frac{\Delta H_c}{\theta \, \Delta H_v} - \frac{H_s}{\Delta H_v} \quad \text{(an enthalpy driving force)} \tag{30b}$$

where

$\dot{m}''$ = burning rate (kg/sec m^2)

g = acceleration of gravity (m/sec^2)

ΔH_c = heat of combustion (HHV) of fuel (kcal/kg)

θ = kilograms oxidant (air) per kilogram fuel (kcal/kg)

ΔH_v = heat of vaporization of fuel (kcal/kg)

H_s = heat content (sensible) of 1 kg of oxidant (air) at the liquid temperature at the surface of the burning pool (kcal/kg)

ρ_a = density of air at ambient temperature (kg/m^3)

and where the following are the properties of the gas mixture formed upon stoichiometric combustion of the fuel and subscripted 1, 2, or 3 to correspond to evaluation at 25%, 50%, and 100%, respectively, of the total enthalpy of a stoichiometric mixture:

λ = thermal conductivity (kcal m^{-1} sec^{-1} °C^{-1})

C_p = specific heat (kcal kg^{-1} °C^{-1})

ρ = density (kg/m^3)

μ = viscosity (kg/m sec^{-1})

α = thermal diffusivity (m^2/sec)

To use this correlation, the temperatures corresponding to various enthalpies are easily determined, using a plot such as Fig. 3 (Chap. 2). Mixture properties at a temperature T are calculated for mole fractions f_i of components with molecular weights MW_i:

$$\rho = \sum_i f_i \rho_i \tag{31}$$

$$\mu = \frac{\sum_i f_i (MW)_i^{1/2} \mu_i}{\sum_i f_i (MW)_i^{1/2}} \tag{32}$$

$$\lambda = \frac{\sum_i f_i (MW)_i^{1/3} \lambda_i}{\sum_i f_i (MW)_i^{1/3}} \tag{33}$$

$$C_p = \frac{\sum_i f_i C_{p_i} (MW)_i}{\sum_i f_i (MW)_i} \tag{34}$$

$$\alpha = \frac{\lambda}{\rho C_p} \tag{35}$$

As the pool diameter increases beyond the 20 to 30 cm range, radiation from the flame plume plays an increasingly important role in setting the evaporation (burning) rate. For these larger pools where the flame plume is optically thick, the burning rate of a liquid with a density ρ_ℓ (g/cm^3), is given by:

$$\dot{m}'' = \frac{0.076\rho_\ell \Delta H_c}{\Delta H_v} \quad \text{kg m}^{-2} \text{ sec}^{-1} \tag{36}$$

where terms are defined as for Eq. (30). An analysis by Steward [20] of the data of several investigators developed a relationship correlating the height of the visible flame plume with the pool size and burning character- istics. Steward's relationship showed that the tip of the visible flame plume corresponded to a condition of 400% excess air.

2. Droplet Burning

For most liquid waste incineration and liquid fuel burning equipment, atomi- zation is used to subdivide the feed for more efficient and complete burning (see Fig. 9). Atomization is effected by a number of means, including

Low pressure (0.03 to 0.35 atm) air atomization

High pressure (2 atm or more) air or steam atomization

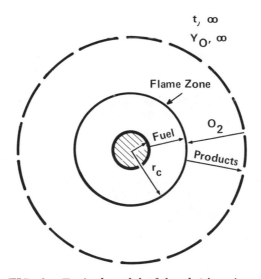

FIG. 9. Typical model of droplet burning.

Mechanical atomization of fluid through special orifices at pressures from 5 to 20 atm

High speed rotating conical metal cups

The power requirements and characteristics of each atomizing method are described below.

Early analysis of droplet burning was based on analogies with coal combustion. These studies viewed droplet combustion as involving the diffusion of oxidant to the droplet surface followed by heterogeneous (two-phase) reaction. Later work led to the presently accepted concept that combustion occurs in a homogeneous diffusion flame surrounding an evaporating droplet (Fig. 9).

Extrapolating from the single droplet model[*] to the analysis of real, large scale combustors requires addition of four rather complex refinements to the theory:

1. In dense sprays, evaporation and diffusion burning of the vapor become the dominant mechanism.

2. Relative motion betweeen droplet and the gaseous surrounding medium removes the simplifying assumptions of radial symmetry.

3. Initial droplet particle sizes are not uniform, and the evaporation characteristics are poorly described using models based on mean droplet sizes.

4. Mixing processes of fuel and oxidant take place in the aerodynamically complex flow in the burner and furnace (not in a simplified, one-dimensional reactor).

Unfortunately, such studies have not yet produced quantitatively useful correlations for use in combustion chamber design. As a starting point, however, the flame length relationship in Eq. (28) could be used for jets of liquid fuel droplets. Application of Eq. (28) is subject to the qualification that the flame will be longer than predicted with the difference in predicted and actual decreasing as

---

[*] A review by Williams [20a] suggests the following as the burning time t (sec) of a droplet of initial diameter d_0 (cm) of a hydrocarbon oil of molecular weight M_W at a temperature T in an atmosphere with an oxygen partial pressure of p_{O_2} (atm) as

$$t = \frac{29800 M_W d_0^2}{p_{O_2} T^{1.75}}$$

The number of coarse droplets decreases

The latent heat of evaporation (less any preheat) decreases

The mean droplet diameter decreases

The flame can be considerably shortened by swirl effects.

C. SOLID COMBUSTION

The combustion of solid phase wastes, as with liquids, is largely a process of gasification followed by combustion of the gaseous products (see Fig. 10). Even in the cases where heterogeneous attack of the solid by a gaseous oxidant such as O_2, CO_2, or H_2O occurs (e.g., in combustion of coke), the product is often carbon monoxide rather than fully oxidized CO_2.

The development of furnaces to burn solids has led to three characteristic firing modes:

1. Suspension burning. Finely subdivided solids, entrained in a fast-moving air stream, are injected into a hot chamber. Combustion is completed while the particles are suspended in the gas flow. Much of the residue leaves the system still entrained in the flue gases.

2. Semisuspension burning. Coarsely subdivided solids are injected mechanically or pneumatically into the combustion space. Drying and combustion occurs partially while the solids are airborne, and is completed after the particles fall to a grate. Residue is dumped periodically or is continuously withdrawn.

3. Mass burning. Unsubdivided or coarsely subdivided solids are moved into the combustion space by mechanical means (e.g., by manual charging or on a grate), are dragged through, tumble through, or remain undisturbed in the combustion space until a satisfactory degree of burnout is obtained. Residue is discharged periodically or continuously.

In waste disposal practice, each of these techniques has been used. Although quantitative descriptions of the processes involved are not yet available, several of the important steps are subject to analysis. A summary of the results are given below to give insight into the processes and their control.

In suspension burning, particles of fuel are exposed to intense radiant heating and are heated convectively and by conduction as hot combustion products are entrained into the jet of suspending fluid. In mass burning, much of the mass is shielded from radiant heating. Thus, fuel particles deep within the bed must await the arrival of the ignition front before rapid

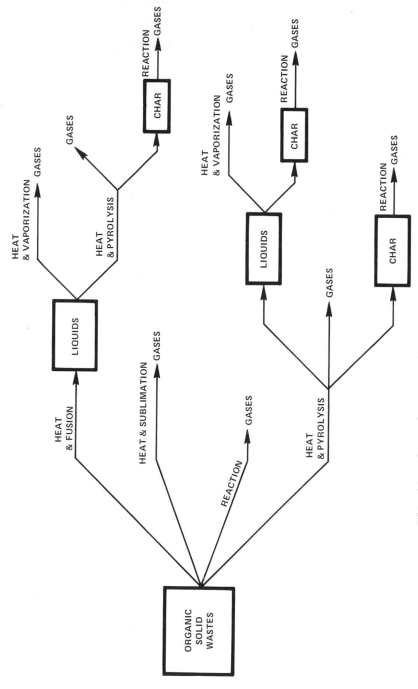

FIG. 10. Phase changes in the combustion of organic solids.

temperature increases are experienced. Semisuspension burning processes fall between the extremes. As the temperature of a fuel particle increases, free moisture is lost, and then decomposition of organic matter beings. The volatile matter includes a wide variety of combustible hydrocarbons, carbon monoxide, carbon dioxide, water vapor, and hydrogen. The weight loss and composition of the volatiles from the solid depend on the time and temperature history. In pulverized coal furnaces, the heating rates are very high ($10^4°$ C/sec or more). In such cases, decomposition is complete in less than 1 sec for particles smaller than about 100 μm.

At the other extreme, a slower rate of heating (less than 10° C/min) characterizes carbonization process in charcoal manufacture. Intermediate heating rates would be found in waste incineration by mass burning (say, 1000° C/min), or the proximate analysis test (about 300° C/min).

The significance of these processes to the incinerator designer arises from their impact on the spatial distribution of volatile release, and thus combustion air requirement within the furnace. If, for example, volatile release is extremely rapid, the region near the charging point of a mass burning system would behave as an intense area source of fuel vapors, driven out of the fuel bed by undergrate air flow. Combustion air requirements, therefore, would be largely in the overbed volume in this region of the furnace and the high, localized heat release rate must be taken into account in developing the design of the enclosure. If, on the other hand, volatilization is slow or limited, heat release will be confined largely to the bed and the majority of the combustion air should be underfire. With a perversity not uncommon in the analysis of waste disposal problems, the intermediate situation appears to be the case.

The introduction of oxidant only in localized areas of the incinerator environment introduces a second complication. In the situation where a mass of ignited solid waste sits upon a grate through which air is passing, the gaseous environment varies with distance above the grate line. In the idealized case, the following processes occur:

> At the grate, air supply is far above stoichiometric, and released volatiles burn rapidly and completely to CO_2 and water vapor. Heat release rates per unit volume are high. One could call this the burning zone.

> At some point away from the grate, the free oxygen is exhausted, and both CO_2 and H_2O act as oxidizing agents to produce CO and H_2 in both heterogeneous and homogeneous reactions. This is the gasification zone.

> Further up in the bed, no oxidizing potential remains in the gases, and the effect is one of immersion of solid waste in a hot gas flow with pyrolysis reactions predominating. This is the pyrolysis zone.

In real situations, the bed is often split open in places due to the heterogeneous form of solid wastes. This leads to channeling or bypassing effects. Also, some air may be entrained into the bed from the overbed space. Further, differences in the point-to-point underfire air rate (m^3/sec m^{-2}) due to grate blinding and spatial variation in the density of combustible throughout any vertical plane in the bed leads to irregularity in volatilization patterns. The net result of these factors is smudging the boundary between zones and production, at the top of the bed, of a gas flow which varies moment to moment from fuel-rich to air-rich.

1. Thermal Decomposition

The thermal decomposition or pyrolysis of waste solids in the absence of air or under limited air supply occurs in most burning systems. Pyrolysis is a destructive distillation process effected by the application of heat in an insufficiency of air to yield gaseous, liquid (after cooling), and solid products. In comparing suspension burning and mass burning processes, both involve pyrolysis of incoming solids, but for suspension burning the physical scale of the fuel-rich zone is smaller, and the pyrolysis products will differ due to the differences in heating rate. Figure 11 illustrates the sequence of steps in the pyrolysis of an organic, char-forming material.

Both physical and chemical changes occur in solids undergoing pyrolysis. The most important physical change in materials such as bituminous coal and some plastics is a softening effect resulting in a plastic mass, followed by resolidification. Cellulosic materials increase in porosity and swell as volatiles are evolved.

As cellulose pyrolysis begins (at about 200°C), complex, partially oxidized tars are evolved. As the temperature increases, these products further decompose or crack, forming simpler, more hydrogen-rich gaseous compounds and solid carbon. The solid residue approaches graphitic carbon in chemical composition and physical structure.

Whether the overall pyrolysis process of a given solid is endothermic or exothermic depends on the ultimate temperature attained. For most materials, the process is endothermic at lower temperatures and exothermic at higher temperatures.

The rate controlling step in pyrolysis (heat transfer rate into the waste or reaction rate) is dependent upon the temperature and the physical dimensions of the waste. Below 500°C, the pyrolysis reactions appear rate controlling for waste pieces less than 1 cm in size. Above 500°C, pyrolysis reactions proceed rapidly, and heat transfer and product diffusion are rate limiting. For large pieces (greater than 5 cm), heat transfer probably dominates for all temperatures of practical interest.

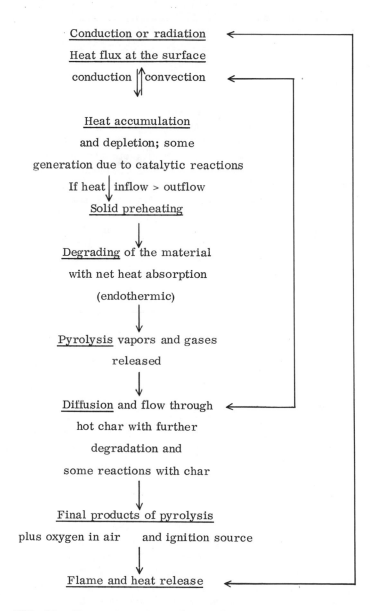

FIG. 11.   Process sequence in pyrolysis.

The upper temperature limit for pyrolysis weight loss is a function of the material being heated. For bituminous coal the weight loss achieved at 950°C is about as high as will be obtained at any practical temperature. Cellulose pyrolysis is essentially completed between 575 to 700°C.

a. Pyrolysis Time

The time required for pyrolysis of solid wastes can be estimated by assuming that the rate is controlled by the rate of heating. Neglecting energy absorption or generation by the reactions, one may estimate pyrolysis time by assuming the refuse piece (a plate or a sphere) is suddenly plunged into a hot reactor maintained at a high temperature. The pyrolysis (heating) time is defined here as the time required for the center temperature to rise by 95% of the initial temperature difference between the specimen and the surroundings. A thermal diffusivity of 3.6×10^{-4} m^2/hr was used; roughly equal to that of paper, wood, coal, and many other carbonaceous solids. Radiation and contact conduction are neglected. The results for thin plates or spheres [21] are shown in Figs. 12 and 13. The effect of specimen size is evident, thus showing the differences expected with size reduction. The effect of gas velocity is also significant and illustrates that in mass burning, where relative gas-solid velocities may range from 0.1 to 0.4 m/sec, slow pyrolysis times would be expected. The relatively long pyrolysis time calculated for small particles at very low (0-0.1 m/sec) relative velocity (as would be found in suspension burning), suggests strongly that other heat transfer mechanisms (especially radiation) can become of primary importance. The results for entirely radiant heating to the plate or sphere starting at an initial temperature of 20°C is also shown in Figs. 12 and 13 as the heating time at infinite cross-flow velocity (V_∞). The indicated heating time requirement is seen to be large compared to the residence time in the radiant section of a boiler (approximately 1-2 sec) and thus, without considerable subdivision, unburned material can enter the convection tube banks.

b. Pyrolysis Products

(1) Liquid Products

The liquid product of pyrolysis is known to contain a complex mixture of alcohols, oils, and tars known as pyroligneous acids. Pyrolysate liquids from municipal refuse contain from 70 yo 80% water with the remainder being various alcohols, ketones, acetic acid, methanol, 2-methyl-1-propanol, 1-pentanol, 3-pentanol, 1,3-propanediol, and 1-hexanol. The exact composition is highly dependent upon refuse composition, heating rate, and ultimate temperature (see Tables 9, 10, and 11). The yield of liquid is approximately 50 to 60% of the air-dried, ash-free refuse, decreasing with increasing ultimate temperature. The heat content of the liquid per pound of refuse decreases as the pyrolysis temperature increases.

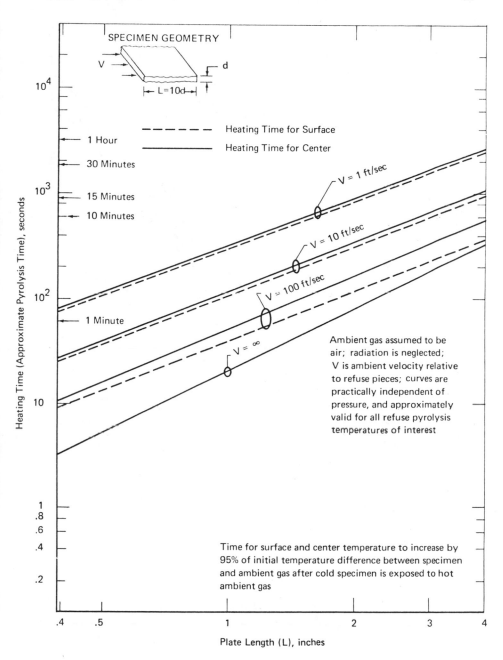

FIG. 12. Radiative and convective heating time for a thin plate (from Ref. 21).

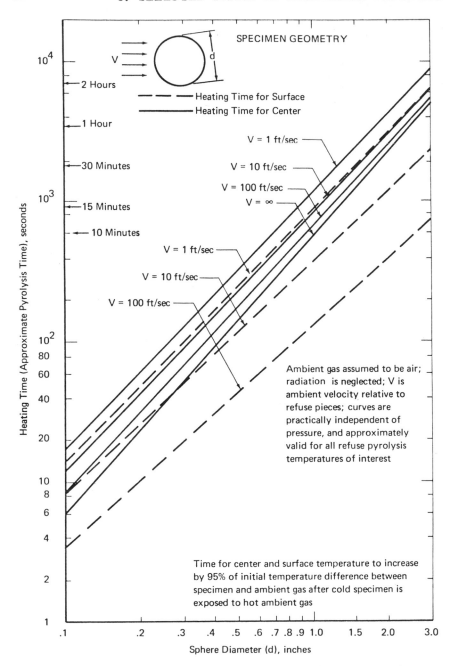

FIG. 13. Radiative and convective heating time for a sphere (from Ref. 21).

TABLE 9

Yields of Pyrolysis Products from Different Refuse Components
(Weight Percent of Refuse)[a]

Component	Gas	Water	Other liquid	Char (Ash-free)	Ash
Cord hardwood	17.30	30.93	20.80	29.54	0.43
Rubber	17.29	3.91	42.45	27.50	8.85
White pine sawdust	20.41	32.78	24.50	22.17	0.14
Balsam spruce	29.98	21.03	28.61	17.31	3.07
Hardwood leaf mixture	22.29	31.87	12.27	29.75	3.82
Newspaper I	25.82	33.92	10.15	28.68	1.43
Newspaper II	29.30	31.36	10.80	27.11	1.43
Corrugated box paper	26.32	35.93	5.79	26.90	5.06
Brown paper	20.89	43.10	2.88	32.12	1.01
Magazine paper I	19.53	25.94	10.84	21.22	22.47
Magazine paper II	21.96	25.91	10.17	19.49	22.47
Lawn grass	26.15	24.73	11.46	31.47	6.19
Citrus fruit waste	31.21	29.99	17.50	18.12	3.18
Vegetable food waste	27.55	27.15	20.24	20.17	4.89
Mean values	24.25	23.50	22.67	24.72	11.30

[a]Refuse was shredded, air-dried, and pyrolyzed in a retort at 815°C. From Ref. 22.

(2) Gaseous Products

The yield, composition, and the calorific value of gaseous pyrolysates depend upon the type of material, the heating rate, and the ultimate temperature. Table 9 shows that for typical refuse components, gas yield ranges from 17 to 31% of the air-dried feed. From Table 10 it can be seen that a doubling in gas yield occurs as the ultimate temperature is raised from 480 to 925°C. Table 11 indicates that the yield decreases, then increases with heating rate.

As the ultimate temperature increases, more of the feed material combustion energy appears in the gas, although the data indicate that the

TABLE 10

Yields of Pyrolysis Products from Refuse at Different Temperatures
(Percent by Weight of Refuse Combustibles)[a]

Temperature (°C)	Gases	Liquid (including water)	Char
480	12.33	61.08	24.71
650	18.64	59.18	21.80
815	23.69	59.67	17.24
925	24.36	58.70	17.67

[a]From Ref. 23.

TABLE 11

Effect of Heating Rate on Yields of Pyrolysis Products and Heating Value
of the Pyrolysis Gas from Newspaper[a]

Time taken to heat to 815°C (min)	Yield, weight percent of air-dried refuse				Heating value of gas, kcal/kg of newspaper
	Gas	Water	Other liquid	Char (Ash-free)	
1	36.35	24.08	19.14	19.10	1136
6	27.11	27.35	25.55	18.56	792
10	24.80	27.41	25.70	20.66	671
21	23.48	28.23	26.23	20.63	607
30	24.30	27.93	24.48	21.86	662
40	24.15	27.13	24.75	22.54	627
60	25.26	33.23	12.00	28.08	739
60	29.85	30.73	9.93	28.06	961
71	31.10	28.28	10.67	28.52	871

[a]From Ref. 22.

gas calorific value does not show a regular variation (Table 12). Table 13 shows an increase in hydrogen and a drop in carbon dioxide as ultimate temperature increases. Table 11 suggests that the heating value of cellulosic materials (represented by newsprint) first decreases, then increases with increasing heating rate. In general, pyrolysis of 1 kg of typical refuse combustibles yields from 0.125 to 0.185 m^3 of gas having a calorific value of about 3000 kcal/m^3; about one-third the calorific value of natural gas.

(3) Solid Products

The solid product or char resulting from pyrolysis is an impure carbon. The proximate analysis of the char is similar to coal, with the rank of the corresponding coal increasing as the ultimate pyrolysis temperature increases. Char formed at 480°C is comparable to certain bituminous coals, whereas pyrolysis at 925°C produces an anthracite-like product. The yield of char ranges from 17 to 32% of the air-dried, ash-free feed, decreasing with both increasing heating rate and ultimate temperature (Table 14). The calorific value of refuse char is around 6600 kcal/kg of air-dried char and decreases slowly as ultimate temperature increases (Table 14).

c. Thermal Decomposition Kinetics

Although the details of the complex chemistry and heat transfer processes controlling pyrolysis reactions in wastes (thick sections, often compounded of several materials, anisotropic and heterogeneous in thermal and

TABLE 12

Calorific Value of Pyrolysis Gases Obtained by
Pyrolyzing Refuse at Different Temperatures[a]

Temperature (°C)	Gas yield per kg of refuse combustibles, (m^3)	Calorific value	
		kcal/m^3 gas	kcal/kg refuse combustibles
480	0.118	2670	316
650	0.173	3346	581
815	0.226	3061	692
925	0.211	3124	661

[a]From Ref. 23.

TABLE 13

Composition of Pyrolysis Gases Obtained by
Pyrolizing Refuse to Different Temperatures[a]

Temperature (°C)	Gas composition, % by volume					
	H_2	CH_4	CO	CO_2	C_2H_4	C_2H_6
480	5.56	12.43	33.50	44.77	0.45	3.03
650	16.58	15.91	30.49	31.78	2.18	3.06
815	28.55	13.73	34.12	20.59	2.24	0.77
925	32.48	10.45	35.25	18.31	2.43	1.07

[a]From Ref. 23.

chemical properties) are not known, pyrolysis has been extensively studied
for wood and synthetic polymers. Much of the general body of literature
concerned with this topic is found under the heading "Fire Research" and
derives its support from an interest in the basic processes which start and
feed (or stop) conflagrations in man-made and natural combustible materials.

Studies by Kanury [24] using an x-ray technique to monitor density
changes during the pyrolysis of wooden cylinders subjected to convective
and radiative heating, provide useful insight into pyrolysis kinetics. As a
consequence of the x-ray technique, Kanury's analysis, summarized below,
is based on density change rather than mass.

Based on convention [25-27], surface pyrolysis of wood may be assumed to follow a first-order Arrhenius-type rate law

$$\frac{d\rho}{dt} = -k_1 \, (\rho - \rho_c) \, \exp\left(\frac{-\Delta E}{RT}\right) \tag{37}$$

where

ρ = the instantaneous density (g/cm) and subscripts c = char,
v = virgin solid

t = time (min)

k_1 = the pre-exponential factor (min^{-1})

ΔE = the activation energy (kcal/kg mol)

R = the universal gas constant (1.987 cal mol^{-1} K^{-1})

T = the absolute temperature (K) and subscript 0 indicates initial

TABLE 14

Comparison of Char Produced by Refuse Pyrolysis with Certain Coals[a]

Proximate analysis of air-dried material	Bituminous coal from lower Freeport Seam	Char resulting from refuse pyrolysis at different temperatures				Pennsylvania anthracite
		480°C	650°C	815°C	925°C	
Volatile matter %	23.84	21.81	15.05	8.13	8.30	8.47
Fixed carbon %	65.36	70.48	70.67	79.05	77.23	76.65
Ash %	8.61	7.71	14.28	12.82	14.47	11.50
kcal/kg of air-dried material	7880	6730	6820	6410	6330	7565

[a]From Ref. 21.

Kanury plotted $(\rho - \rho_c)^{-1}(d\rho/dt)$ versus $1/T$, on semilog paper to test the validity of Eq. (37), and found good agreement with an activation energy (ΔE) of about 19,000 kcal/kg mol and a pre-exponential frequency factor of about 10^6 min^{-1}.

Observation of the surface temperature with a microthermocouple indicated an approximately linear change with time:

$$T = T_0 + a_1 t \tag{38}$$

Combining Eqs. (37) and (38) and integrating yields

$$\ln\left[\frac{\rho - \rho_c}{\rho_v - \rho_c}\right] = \frac{-k_1 \Delta E}{a_1 R}\left[\gamma_0(\zeta) - \gamma_0(\zeta_0)\right] + \int_{\zeta_0}^{\zeta} \gamma_0(\zeta)\ d\zeta \tag{39}$$

where

$$\zeta = \frac{\Delta E}{RT}$$

$$\gamma_0(\zeta) \equiv \frac{1}{\zeta}\ \exp(-\zeta)$$

Integration of $\gamma_0(\zeta)$ yields

$$\int \gamma_0(\zeta)\ d\zeta = \ln \zeta + \sum_{n=1}^{\infty} \frac{(-1)^n \zeta^n}{(n)(n!)} \tag{40}$$

The right hand side of Eq. (40) converges for positive ζ and the exponential integral of the first order $\gamma_1(\zeta)$ is defined as

$$\gamma_1(\zeta) = -0.5772 - \left(\ln \zeta + \sum_{n=1}^{\infty} \frac{(-1)^n \zeta^n}{(n)(n!)}\right)$$

$$= \int_{\zeta}^{\infty} \gamma_0(\zeta)\ d\zeta$$

Therefore,

$$\int_0^\infty \gamma_0(\zeta) \, d\zeta = \gamma_1(\zeta_0) - \gamma_1(\zeta) \tag{41}$$

Substituting Eq. (41) into Eq. (39):

$$\ln\left(\frac{\rho - \rho_c}{\rho_v - \rho_c}\right) = \frac{-k_1 \, \Delta E}{a_1 R} \left[\gamma_0(\zeta) - \gamma_0(\zeta_0) + \gamma_1(\zeta_0) - \gamma_1(\zeta)\right]$$

Since ζ_0 is usually $\gg 30$, $\gamma_0(\zeta_0)$ and $\gamma_1(\zeta_0)$ are of the order of 10^{-10} and may be neglected yielding

$$\ln\left(\frac{\rho - \rho_c}{\rho_v - \rho_c}\right) \approx \frac{-k_1 \, \Delta E}{a_1 R} \; \gamma_0(\zeta) - \gamma_1(\zeta) \tag{42}$$

$\gamma_0(\zeta)$ and $\gamma_1(\zeta)$ are tabulated in standard texts.

Kanury's experimental data on the surface density-time function was in excellent agreement with a prediction based on the experimentally determined value for a_1 of 50 K/min.

Kanury's data also showed two plateuas in the pyrolysis rate: at 100 and 350°C. The plateau at 100°C is, most likely, associated with supply of the latent heat of evaporation of moisture in the wood. The plateau at 350°C is suspected to be due to the resistance offered by the char layer to mass flow and diffusion of the pyrolysis gases. It was noted, in support of this hypothesis, that the 350°C plateau was absent in pyrolysis of the surface layers or when major cracks and fissures appeared as the solid near the axis of the cylinder pyrolyzed.

Kanury also observed that if the pyrolysis rate becomes constant (in the plateau region), then from Eq. (37)

$$\frac{d\rho}{dt} = \text{constant} = -a_2(\rho_v - \rho_c)$$

A plot of the data for $\Delta E/R = 9500$ K gave satisfactory agreement for

$$\left(\frac{\rho - \rho_c}{\rho_v - \rho_c}\right) = \frac{a_2}{k_1} \exp\left(\frac{\Delta E}{RT}\right) = 1.436 \times 10^{-7} \exp\left(\frac{\Delta E}{RT}\right) \tag{43}$$

The slope of the plot of the log of the l.h.s. of Eq. (43) versus $1/T$ K had a slope of unity, thus providing additional confirmation of the first order reaction rate expression.

Another observation was that the pyrolysis rate, after attaining its maximum value, fell in a manner somewhat like an exponential, viz.

$$\left(\frac{\rho - \rho_c}{\rho_v - \rho_c}\right) \frac{d\rho}{dt} = -a_3 \exp(-a_4 t) \tag{44}$$

Integration yields

$$\frac{\rho - \rho_c}{\rho_v - \rho_c} = \frac{a_3}{a_4} \exp(-a_4 t) + a_5 \tag{45}$$

Since, as t becomes large, the exponential tends to zero and ρ approaches ρ_c, a_5 is 0. Plotting the data for the pyrolysis rate in the inner layers $a_3 = 1.404$ min^{-1} and $a_4 = 0.216$ min^{-1} or

$$\frac{\rho - \rho_c}{\rho_v - \rho_c} \cong 6.5 \exp(-0.216t) \tag{46}$$

Eliminating the exponential between Eqs. (44) and (46), and substituting into Eq. (37) yields:

$$k_1 \exp\left(\frac{-\Delta E}{RT}\right) \approx 0.216 \quad \text{a constant} \tag{47}$$

For $k_1 = 10^6$ min^{-1} and $\Delta E/R = 9500$ K, $T = 619$ K or $346°$C, the observed plateau temperature. In summary, then, it would appear that the wood cellulose pyrolysis process consists first of an essentially nonreacting thermal heating stage with moisture loss to about $350°$C. Then, pyrolysis occurs at a rapid rate, followed by further heating of the char residue through conduction and radiation from the exposed surface.

Data by Shivadev and Emmons on thermal degradation of paper [28] also showed a plateau in a temperature range about $350°$C. Their experiments involved radiant heating of filter paper samples. They deduced reaction rate constants and energy release terms by radiating preweighed specimens for progressively longer time periods and then determining the total weight loss as a function of time. Temperatures in the specimen were evaluated by considering the specimens to be thermally thin and thus rapidly attaining an equilibrium temperature related to the radiative flux level, the absorptivity of the surface, and material properties, all of which were either known or measured.

Shivadev and Emmons fitted their mass-loss (surface density m) and temperature data with two expressions. The first, for temperatures below 655 K (382°C) is

$$\frac{dm}{dt} = -5.9 \times 10^6 \exp\left(\frac{-26,000}{RT}\right) \quad \text{g cm}^{-2} \text{ sec}^{-1} \tag{48a}$$

For temperatures above 655 K they found

$$\frac{dm}{dt} = -1.9 \times 10^{16} \exp\left(\frac{-54,000}{RT}\right) \quad \text{g cm}^{-2} \text{ sec}^{-1} \tag{48b}$$

They speculate that the expression for the high temperature reaction may incorporate combustion effects of the pyrolysis gases. It is noteworthy here that in associated studies of ignition, the inflection temperature associated with ignition was 680 ± 15 K. The numerical values obtained are in rough agreement with the Kanury data.

2. Mass Burning Processes

In the idealized mass burning model described above it was postulated that in thick beds the upper regions could behave as a true pyrolyzer. Evidence [29, 30] for coal and refuse beds would tend either to discount the existence of the pyrolysis zone or, more probably, to suggest that this zone does not appear in beds of practical thickness. Thus, the off-gas from a bed would be a mixture of gases from both burning and gasification zones.

Recent studies [30] on the off-gas from refuse burning in a municipal incinerator have confirmed the earlier data of Kaiser [31] and the hypothesis [21] that the off-gas composition is controlled by the water-gas shift equilibrium. This equilibrium describes the relative concentration of reactants in the following:

$$H_2O + CO \rightleftharpoons CO_2 + H_2$$

$$K_p = \frac{p_{H_2O} p_{CO}}{p_{CP_2} p_{H_2}} \tag{49}$$

The importance of this equilibrium in mass burning is the incremental gasification potential given to the underfire air. In tests at Newton, Massachusetts [30], refuse and off-gas stoichiometry were studied. Average refuse was given the mole ratio formula: $CH_{1.585}O_{0.625}(H_2O)_{0.655}$. For this formula and assuming that the water-gas shift equilibrium holds, over 1.5 times as much refuse can be gasified as would be predicted for stoichiometric combustion to CO_2 and H_2O. Burning rate data showed rates 1.7 to 2.1 times that corresponding to stoichiometric combustion.

A second result coming from the water-gas shift reaction is that a definite and relatively large combustion air requirement will necessarily be placed on the overfire volume. The air requirement for the CO, H_2, distilled tars, and light hydrocarbons can, indeed, be as much as 30 to 40% of that expended in gasification, thus creating a need for effective overfire air injection and mixing.

D. AIR POLLUTANT GENERATION
    IN COMBUSTION PROCESSES

In the combustion of fuels and waste streams, air contaminants are generated which have significance to the design engineer or system analyst in three areas:

1. Obtaining permits from regulatory agencies for installation and/or operation of the system

2. Establishing the need and/or specifications for air pollution control systems

3. Establishing the basic design, suggesting modifications to existing designs, or interpreting problems related to emission minimization or control

The air pollutants emitted from combustion processes include inorganic particulate matter; combustible solids, liquids, and gases [carbonaceous soot and char; carbon monoxide; "hydrocarbons"; and specialized classes of carbon-hydrogen-oxygen-nitrogen-halogen compounds such as benzene-soluble organics (BSO), polycyclic organic matter (POM), polyhalogenated hydrocarbons (PHH), including polychlorinated biphenyls (PCB), polybrominated biphenyls (PBB), and benzo-(a)-pyrenes (BaP), all of which have or are suspected to have significant carcinogenic (cancer causing) or other health-related importance]; specific chemical pollutants where emission levels are largely related to fuel chemistry (sulfur oxides, chlorine and hydrogen chloride, trace elements, radioactive elements); and nitrogen oxides where emission levels are related to a wide variety of combustor- and fuel-related interactions.

The sections that follow yield some insight as to the mechanisms by which these pollutants arise and thus, to a point, the means available to the designer and operator (both have a role in most cases) to minimize their emission rate.

1. Inorganic Particulate

Inorganic particulate is differentiated from total particulate as an acknowl-
edgment of its refractory nature (i.e., emission rates cannot be reduced by
better combustion), and to clearly establish its primary source, the ash
content of the fuel or waste which is burned. In principle, inorganic particulate can arise from mechanical degradation of refractory, or oxidation
and flaking of fireside metal surfaces, but the great majority of the emission results from the carryover of mineral matter introduced with the fuel
or waste. In unusual conditions, where combustion air is drawn from an
area with a high dust loading, a portion of the emission could be associated
with the air supply.

a. Gaseous and Liquid Wastes and Fuels

A desk top assessment of the fraction of the inorganic matter in a fuel
which appears in the flue gas is often difficult because of the complexity of
the processes involved. Thus, empirical estimation methods are necessary
unless the system design is such that essentially all solids which are introduced must leave via the flue gas stream. This situation is often true for
systems burning gases and liquids. For conventional gaseous and liquid
fuels, combustion systems seldom include air pollution control designed
for particulate control. When burning liquid or gaseous waste streams,
however, sampling or estimation of inorganic content is an appropriate
design step to assure that no control system is needed. Then, consideration can be given to either precombustion removal of the solid matter or to
the installation of a flue gas particulate air pollution control device.

Example 12. A liquid waste stream is to be incinerated and in preparing
the permit a question has been raised as to the need for air pollution control. It is known that 95% of the inorganic suspended solids can be removed
by use of a centrifuge. The applicable particulate emission code sets an
upper limits of 229,000 $\mu g/m^3$, corrected to 20°C, 1 atm, and a (dry) flue
gas CO_2 content of 12% (by volume). Evaluate the situation.

Waste Composition

Component	Weight percent
n-Hexane (C_6H_{14})	46.0
Ethanol (C_2H_5OH)	23.0
Water (H_2O)	28.7
Suspended mineral solids	2.3
	100.0

Basis: 100 kg of waste

Component	kg	Moles	Moles C	Moles H_2	Moles O_2	Moles stoichiometric O_2 required
C_6H_{14}	46.0	0.535	3.209	3.744	0	5.081
C_2H_5OH	23.0	0.5	1.000	1.500	0.25	1.500
H_2O	28.7	1.594	0	1.594	0.797	0.0
Ash	2.3	—	—	—	—	—
Total	100.0	2.629	4.209	6.838	1.047	6.581

In the dry flue gases at test conditions of 12% CO_2

$$12.00 = \frac{(\text{moles } CO_2)(100)}{\text{moles } CO_2 + \text{moles } O_2 \text{ in excess air} + \text{moles } N_2 \text{ in stoichiometric and excess air}}$$

For X moles O_2 in the flue gases

$$12 = \frac{(4.209)(100)}{4.209 + X + (79/21)(6.581 + X)}$$

X = 1.283 moles O_2 (equivalent to 19.5% excess air)

Dry flue gas contains

$$\begin{array}{lll}
CO_2 & 4.209 \text{ mol or} & 12.0\% \\
O_2 & 1.283 \text{ mol or} & 36.6\% \\
N_2 & \underline{29.584} \text{ mol or} & 84.4\% \\
\text{Total} & 35.076 \text{ mol or} & 100\%
\end{array}$$

At code conditions, the dry flue gas will have a volume of

$$(35.076)(22.4)\left(\frac{293}{273}\right) = 843 \text{ m}^3$$

and will contain 2300×10^6 μg of solids for a corrected particulate loading of 2.73×10^6 $\mu g/m^3$, in excess of the code. Ninety-five percent suspended solids removal will reduce the loading of mineral particulate to 136,500 $\mu g/m^3$, or about 60% of the code. The decision is thus clear: either waste pretreatment or air pollution control will be required, with the selection of the optimum to be based on a comprehensive economic analysis (including capital and operating costs, centrifuge solids or ash disposal costs, etc.)

b. Solid Wastes and Fuels

Unlike liquid or gaseous fuels which may contain little inorganic matter, almost all solid wastes and fuels have a substantial ash content. Since
these latter fuels are seldom susceptible to pretreatment to reduce emissions, the designer must give careful attention to design features and operating practices to minimize the emission rate. Even with such precautions,
it is likely that emission control will be required.

(1) Suspension Burning

The maximum inorganic particulate emission rate corresponds to the
situation where all noncombustible solids are swept from the combustion
chamber. This is largely the case for burners fired with pulverized coal,
shredded and air-separated municipal refuse, sawdust, rice hulls, and
other suspension firing systems. In such systems, a portion of the fly ash
may settle in the bottom of the primary furnace or in following chambers or
boiler sections. The degree to which settlement reduces emission rate
depends on the flue gas properties, on the characteristic dimensions and
weight of the fly ash particles (settling rate), and on the velocity of the gas
stream relative to the chamber dimensions (transit time).

An approach to estimating the potential for fallout makes use of Stokes'
Law (and its various extensions into the "transition" regime for larger particles), although the result will tend to overestimate the fallout rate. The
complications tending to lower the measured settling efficiency below the
calculated value are

Nonsphericity of particles.

Particle reentrainment due to turbulence. This can be minimized
by use of baffles (creating low turbulence zones or, if appropriate, a
wet, water-sluiced chamber floor.

Nonuniformity of particle density: the coarser particles tend to be
lower in density.

Particle-particle interactions ("hindered settling").

Fallout of the very coarse particles prior to the chamber under
analysis leaving a less easily settled dust in the gas stream.

In general, settling velocities for homogeneous spherical particles can
be calculated from the relation

$$u_t = \sqrt{\frac{4d(\rho_s - \rho_a)g}{3\rho_a C_x}} \tag{50}$$

where

u_t = terminal settling velocity, m/sec

d = particle diameter, m

ρ_s = density of particle, kg/m^3

ρ_a = density of gas, kg/m^3

g = gravitational constant, m/sec^2

C_x = drag coefficient, dimensionless

In the streamline flow region (small particles), the drag coefficient is inversely proportional to the Reynolds number N_{Re}:

$$C_x = \frac{24}{N_{Re}} = \frac{24\mu}{du_t\rho_a} \tag{51}$$

where

μ = viscosity of gas, kg/m sec^{-1}

N_{Re} = Reynolds number = $\dfrac{(du_t\rho_a)}{\mu}$ which is dimensionless

In the streamline flow region, combination of Eqs. (50) and (51) leads to Stokes' Law:

$$u_t = \frac{d(\rho_s - \rho_a)g}{18\mu} \tag{52}$$

For practical purposes, the streamline flow region stops at about Re = 0.5 to 1.0. Thus, for 200°C gases and particulate densities of 2.5 g/cm^3 (2500 kg/m^3), Eq. (51), and thus Stokes' Law, is valid only up to diameters of about 90 μm. Above this region, we must use relationships other than Eq. (51) to characterize the dependence of C_x on Re. For larger particles, the relation of Schiller and Naumann [32] indicates

$$C_x = \frac{24}{Re} [1 + 0.15 \, Re^{0.687}] \tag{53}$$

This relation is valid for 0.5 < Re < 800.

Unlike the combination of Eqs. (50) and (51), the combination of Eqs. (50) and (53) no longer gives an explicit relation for u_t, but rather requires a trial and error solution. As an example, application of these relations to particles of density 2.0 and 3.0 g/cm^3 in hot air at two temperatures leads to the settling velocities shown in Table 15. Note that for large particles, the effect of temperature is not very great.

TABLE 15

Calculated Settling Velocities in Hot Air
(Spherical Particles)[a]

Diameter (μm)	Settling velocity (m/sec)			
	(Density = 2.0 g/cm^3)		(Density = 3.0 g/cm^3)	
	at 200°C	at 800°C	at 200°C	at 800°C
30	0.03	0.03	0.06	0.03
40	0.06	0.03	0.09	0.06
50	0.12	0.06	0.15	0.09
60	0.15	0.09	0.24	0.12
70	0.21	0.12	0.34	0.18
80	0.27	0.15	0.43*	0.24
90	0.37*	0.21	0.49	0.30
100	0.43	0.24*	0.55	0.37*
200	1.13	0.85	1.55	1.22
400	1.92	1.65	2.56	2.29
400	2.68	2.50	3.54	3.41
500	3.38	3.35	4.45	4.54
600	4.08	4.21	5.33	5.64
700	4.75	5.03	6.16	6.71
800	5.36	5.85	6.98	7.74
900	6.00	6.64	7.89	8.75
100	6.58	7.41	8.50	9.72

[a] The asterisk denotes the largest particle for which Stokes' Law was used; Schiller-Naumann extension [32] was used for all larger particles.

For a chamber 9 m high with a mean gas residence time of 5 sec (similar to the settling chamber originally installed at the Hamilton Avenue incinerator in New York City for operation at 300% excess air), all particles falling faster than about 1.8 m/sec (or bigger than about 250 μm in diameter, from Table 15) should reach the bottom of the chamber during the 5-sec transit time. Particles falling at a slower rate should be deposited in proportion to the ratio of their settling velocities to the critical velocity (1.8 m/sec for this chamber). The performance for a nominal size distribution [21] would be as shown in Table 16. Such a "perfect" chamber would remove about 41% of the incoming particulate weight. In practice, due to the factors mentioned above, a typical efficiency for a dry settling chamber is only 15% [33]. In Installation No. 1 as described by Walker and Schmitz [34], the efficiency of such a chamber is estimated at 21%. Operation with wet floors to prevent reentrainment can increase the efficiency, for example, to measured values of about 35% at Hamilton Avenue and to 31% at another installation. A later test at the latter plant gave a 21% efficiency. Baffling can also raise the efficiency by increasing the probability of a particle encountering a solid surface.

The theoretical performance of other settling chambers can be calculated as above, with the aid of the settling velocities given in Table 15. Thus, for example, a shorter drop—say a 4.5 m height with the same residence time at 200°C—would, in theory, remove all particles larger than 150 μm and have a total removal efficiency of about 47% (compared with the 41% figure for the chamber twice as high). The efficiency of these devices is low, but so is their pressure drop—typically below 0.3 mm Hg. The major portions of this pressure loss occur at the inlet and a lesser amount at the outlet of the chamber, as a result of expansion and contraction losses.

(2) Mass Burning

In mass-burning, the fuel or waste is moved through the combustion chamber on a grate. The introduction of a portion of the combustion air through the grate provides a mechanism whereby a portion of the ash can be fluidized and carried off with the flue gases. This mechanism is favored by fuels or wastes with a high percentage of fine (i.e., suspendable) ash, by high underfire air velocities, or by other factors that induce a high gas velocity through and over the bed. Secondary inducements to emission include agitation of the bed and the volatilization of metallic salts. Grate systems designed with large air passages (such that a substantial fraction of the fine ash is dropped out) tend to reduce the emission rate.

(a) Ash Content. Ash particles may be entrained when the velocity of the gases through the fuel bed exceeds the terminal velocity of the particles as calculated using Stokes' Law [Eq. (52)]. Undergrate air velocities in municipal incinerators typically vary from a minimum of 0.05 sm^3 sec^{-1} m^{-2} of

TABLE 16

Calculated Performance of a Settling Chamber

(Nominal dust size and density [21]: 9 m high, 5-sec residence time)

Diameter (μm)	Weight (kg) of particles per 100 kg of total dust entering	Removal efficiency (%)[a]	Weight (kg) of particles remaining	Weight (kg) of particles per 100 kg of total dust remaining
<1	9	0	9	15
1–2	5	0	5	9
2–5	7	0	7	12
5–10	6	0	6	10
10–15	5	0	5	8
15–20	3	0	3	5
20–30	5	1	5	8
30–40	3	3	3	5
40–50	4	5	4	7
50–100	8	15	7	12
100–200	8	50	4	7
200–300	5	90	1	2
>300	32	100	0	0
Total	100		59	100

[a]Removal efficiency = 100−59 = 41%. Based on settling velocities given in Table 15.

grate area to 0.5 sm^{3}sec^{-1} m^{-2}. Based on the terminal velocity of ash particles (Table 15) it is, therefore, expected that particles up to 70 μm will be entrained at the lowest velocities and up to 400 μm at the highest at a mean temperature of 1100°C. The postulated mechanism of particulate entrainment is supported by the observed range of particle sizes of fly ash, the increases in particulate emission with increases in refuse ash content, the increases in particulate emission with undergrate air velocity, and the similarity in the chemical analyses of fly ash and those of the ash of the principal constituents of refuse.

Evaluation of the particle size distribution of fly ash [21] shows 70 wt % of the particles are smaller than 250 μm. This is consistent with the calculations of terminal velocities. The data available, however, are insufficient to test the expectation that the maximum particle size of the fly ash emitted increases with undergrate air velocity.

The dependence of the particulate emission factors on the ash content of refuse is most striking. The very high ash content of refuse in Germany, particularly during the winter in areas where a high fraction of the residential furnaces are coal-fired, accounts for their unusually high emission rates. The correlation of furnace emission factors with ash content of refuse is demonstrated by the data abstracted in Table 17 from articles by Eberhardt and Mayer [35] and by Nowak [36]. The refuse ash content and furnace emission rates in January are greater than the corresponding values in August by factors exceeding 4 and 6, respectively. The percentage of the refuse ash carried over would be smaller if the value were based on the fine particle fraction of the ash in place of the total ash content, but the percentage of the fine ash carried over would, of course, be larger. It is interesting to note that values reported for the particulate emission for a chain grate stoker burning coal corresponds to a little over 20% of the ash content of the coal, a value not very different from the range found for refuse.

Additional evidence of the effect of ash content on particulate is provided by the Public Health Service (PHS) studies on an experimental incinerator [37]. Tests were conducted first with a synthetic refuse consisting of newspaper, cardboard, wood, and potatoes and then repeated with a high-ash paper substituted for the newspaper. The ash contents have been here estimated to be 2.0 wt % for a 25% moisture synthetic refuse and 7.5 wt % for the case in which high-ash paper is substituted for the refuse. These estimates were obtained from the percentages of the different components reported by the PHS and from the ultimate composition of typical refuse components reported by Kaiser et al. [38-40]. The results, summarized in Table 18, show that, for large percentages of underfire air, the percentage of the refuse ash carried over in the combustion gases is approximately the same for the two different fuels and ranges from 8 to 22% with a maximum value a little greater than that found in European

TABLE 17

Seasonal Variation in Refuse Ash Content
and Furnace Emission Rate[a]

Month	Refuse ash content (%)	Refuse heating value (kcal/kg)	Furnace emission rate (g/Nm^3)[b]	Refuse ash emitted (%)
January	44.5	1200	15	14
February	38	1311	12.8	15
March	30.5	1422	9.3	13
April	27	1467	7	13
May	22	1533	5.3	12
June	18	1650	4	13
July	12.5	1667	2.3	10
August	10.5	1583	1.9	10
September	14.5	1578	2.7	10
October	26	1444	5.7	11
November	35.5	1350	10	13
December	43	1289	13.7	14

[a] From Refs. 35 and 36.
[b] Grams of particulate per normal (volume of a gas at standard conditions of temperature and pressure) cubic meter.

practice. However, at low underfire air rates, the percentage of the fly ash carried over is much smaller, and better agreement is found between the absolute emission rates for low and high ash content fuels than between the corresponding percentages of ash carried over.

These findings show that the ash content of the refuse is a major factor in determining emission rates, and that the percentage of the ash carried over ranges mostly between 10 and 20% of the total. The actual percentage carried over for a particular incinerator will, of course, depend on other factors, such as the underfire air rate. Consideration of the particle terminal velocities indicates that the above conclusions apply only to fine particle ash (finer than 400 μm).

TABLE 18

Effect of Ash Content on Emission Rates[a]

Refuse ash content (%)	Underfire air (% of total)	Emission rate kg/tonne	Refuse ash emitted (%)
2.0	15	0.57-2.19	2.74-11
2.0	60	1.55-4.46	7.75-22
7.5	15	0.74-1.86	1.0-2.5
7.5	60	6.61-15.83	8.8-21

[a]From Ref. 37.

The above conclusion, that approximately 10 to 20% of the fine ash content of refuse is carried over, may be tested by application to a municipal refuse composition. A tonne of typical refuse, as received, contains 54 kg of fine ash (excluding the metal and glass components) and thus would be expected to yield a furnace emission of 5.5 to 11 kg/tonne. This is in fair agreement with the middle range of emission rates for incinerators in the United States. In addition, the composition of the fly ash may be calculated from the composition of the ashes of the refuse constituents, assuming that no selective entrainment occurs. The composition of fly ash so estimated is compared in Table 19 with analyses of the combustible free fly ash collected in three New York City incinerators [40] and the average analysis of 25 slag samples obtained from nine incinerators in New York, New Jersey, and Connecticut [39]. The agreement between the data on fly ash composition and the values computed from the ash of a typical refuse is satisfactory, with the major differences being the higher values of iron, aluminum, and sulfur oxide found in the ash collected in the New York City incinerators. Oxidation of most of the aluminum foil and part of the iron refuse is the possible explanation for the high percentage of aluminum and iron oxides in the fly ash. The high sulfur content may be attributed to calcium sulfate in wallboard or plaster. From these results, the ash entrainment mechanism seems to be the dominant one in municipal incinerators.

It should be noted that the upper limit given above for the size of particle that could be entrained corresponds to the diameter of a spherical particle. Much larger, thin, flat platelets can be entrained; for example, charred paper bits, known as "blackbirds." Unfortunately, most sampling technicians do not collect these large elements, and no quantitative measure of their rate of emission is available.

TABLE 19

Composition (Weight Percent) of Inorganic Components of Fly Ash

| Component | Computed for typical refuse | NYC incinerators[a] | | Average analysis of 25 slag samples[b] |
		73rd St., Manhattan	So. Shore, Long Island	
SiO_2	53.0	46.4	55.5	44.73
Al_2O_3	6.2	28.2	20.5	17.44
Fe_2O_3	2.6	7.1	6.0	9.26
CaO	14.8	10.6	7.8	10.52
MgO	9.3	2.9	1.9	2.1
Na_2O	4.3	3.0	7.0	8.14
K_2O	3.5	2.3	—	—
TiO_2	4.2	3.1	—	—
SO_3	0.1	2.7	2.3	3.69
P_2O_5	1.5	—	—	1.52
ZnO	0.4	—	—	1.54
BaO	0.1	—	—	—
	100.0			

[a] From Ref. 40.
[b] From Ref. 39.

(b) Underfire Air Rate. A systematic study of the effects of underfire air, secondary air, excess air, charging rate, stoking interval, and fuel moisture content on the emission rate from an experimental incinerator by the PHS led to the conclusion that the velocity of the underfire air was the variable that most strongly influenced particulate emission rate. The data on 25 and 50% moisture fuel were correlated [37] by

$$W = 4.35V^{0.543} \tag{54}$$

where W is the emission factor expressed in units of kilograms of particulate per ton of refuse burned, and V is the underfire air rate (in $sm^3 \ sec^{-1} \ m^{-2}$ of grate area). The range of undergrate velocities studied ranged from

0.01 to 0.5 sm^3 sec^{-1} m^{-2} of grate area. Subsequent field evaluation by the
PHS [41] of emission rates indicate that the effect of underfire air rate was
less pronounced than that predicted by the above equation and suggested that,
for the two municipal incinerators tested, the effect was small for undergrate velocities below 0.18 to 0.2 sm^3 sec^{-1} m^{-2} of grate area but significant
above that rate.

Walker and Schmitz [34], however, correlated their test results on
three municipal incinerators with the relationships developed by the PHS on
the experimental incinerator. The results showed general agreement with
the predicted slope but scattered ±20%. This is not surprising in view of
probable differences in refuse composition, furnace size, and other variables. Although the data demonstrating the increase of particulate emission
with underfire air is limited, supporting evidence for the trend observed is
provided by numerous observations of this effect in the literature [42-45].

Additional data supporting the conclusions derived in the literature are
those on Plant No. 76 [21]. The tests of this furnace showed that the emission increased 59% as the underfire air rate was increased from 50 to 100%
of stoichiometric, but that decreasing the overall excess air from 50 to
20% resulted in a 44% increase in emission. The anomalous behavior at
20% excess may be attributed to very poor burning characteristic and extensive hand stoking required at the low excess air limit.

Although reduction in the underfire air rate can reduce the emission
from the furnace, there is usually [41,43] an attendant reduction in burning
rate. An economic analysis of the trade-off between furnace capacity and
air pollution control equipment costs would be needed to optimize the underfire air rate. There is, of course, a minimum underfire air requirement
necessary to protect the grate.

(c) Incinerator Size. Larger incinerator units seem to have slightly higher emission rates, but the effect of size on emission factors has not been
quantitatively established. Part of the increase is due to the higher burning
rates and, hence, higher underfire air rates associated with large units.
However, if emission on large and small units is compared at equal underfire air velocities, a residual effect of size is found. Possibly the higher
emission rates for the larger size is a consequence of the higher natural
convection currents encountered in large units (see page 186, ff.).

(d) Burning Rate. For reasons similar to those presented in the preceding
paragraph, it is expected that higher emission factors will be encountered
at higher burning rates. Rehm [46] cites that reductions in rate of burning
to 75% of rated capacity have shown as much as 30% reduction in furnace
emission from that at full capacity; however, insufficient data are available
for a quantitative relationship to be established.

(e) <u>Grate Type</u>. Stoking has been observed to increase particulate emission rates. This is particularly evident from the tests on Plant No. 76 [21] and the PHS field tests [41] on the effect of underfire air rate, where the stoking required at the lowest air rates led to a significant increase in emission. Walker and Schmitz [34], however, report results that suggest that the effect of grate design on emission is secondary to the effect of underfire air rate or ash content. The results of their tests on the emission from furnaces operated with a two-section traveling grate, a rocking grate, and a reciprocating grate led them to conclude that the differences in emission factors were primarily due to the differences in underfire air velocities used in the different units.

In order to assess the effect of grate type on emission factors, data on the different types of units were compiled [21]. The emission rates from reciprocating grate stokers were seen to be significantly higher than those from other grate types. It is probable that higher rates in the reciprocating grate units are due to a combination of greater stoking, higher underfire air rates, and larger furnace sizes than in the other units. Another factor is the difference in grate openings, which can result in large differences in the amount of fine ash that can sift through the grate; lower emission rates would be expected from grates (e.g., rocking grates) that have large amounts of fine ash sifting through the grate.

(f) <u>Combustion Chamber Design</u>. Practically the only data in this area are provided by the studies in Los Angeles County. From these data it can be concluded [43] that the emission from small units can be reduced substantially by use of multichambers and by use of a low arch. The emission in these tests was primarily particles under 5 μm in size, and there is uncertainty concerning the applicability of the results to large municipal incinerators. The need for a secondary combustion chamber on large units will depend mainly on the adequacy of mixing in the primary chamber.

(g) <u>Volatilization of Metallic Salts</u>. Although a major fraction of fly ash from municipal incinerators seems to have been entrained from the fuel bed or formed by cracking of pyrolysis products (soot), trends in the emission rate from small incinerators cannot be explained solely by an entrainment mechanism [43,47]. For example, the sizes of the particulate in the stack discharge data reported by Rose and Crabaugh [43] are mostly in the 0 to 5 μm range. Based on the physical shape as determined from microscopic examination, and on chemical analyses of the noncombustible, they concluded that particles were formed by volatilization and recondensation of metallic salts. This finding, however, is inconsistent with the data on large units for which particles of 0 to 5 μm usually constitute a small fraction of the total emission. In municipal incinerators, the oxidation of the metals is known to be significant, but the volatilization of the salts formed is only of importance for trace constituents; the majority of the oxides are mechanically entrained from the bed.

2. Combustible Solids, Liquids, and Gases

a. Pollutant Characterization

The incomplete combustion of fuels containing carbon can result in the formation of a wide spectrum of chemical species. The simplest, carbon itself, can contribute importantly to the opacity of the effluent, due both to the refractive index and color of the particles, and to the typically small particle size (thus increasing the light scattering power for a given mass loading).

A second class of pollutants includes the carbon-hydrogen compounds. The chemical nature of these compounds range from methane, ethane, acetylene, and other straight and branch-chained aliphatic compounds to complex saturated and unsaturated ring compounds.

The third class of pollutants includes the carbon-hydrogen-oxygen compounds. These also range from simple compounds such as carbon monoxide and formaldehyde to complex organic acids, esters, alcohols, ethers, aldehydes, ketones, and so forth.

The fourth class of pollutants includes the carbon-hydrogen-nitrogen compounds. These include the amines, N-ring compounds, and other chemical species, some incorporating oxygen.

The importance of these pollutants is associated with almost all criteria of air quality:

> Some are solid particulates and contribute to atmospheric haze and solids fallout.

> Some are photochemically reactive and thus participate in the reactions leading to smog.

> Some are recognized as injurious to plants (e.g., ethylene), animal life respiration (e.g., carbon monoxide) or are known to cause cancer in human beings (e.g., benz-(a)-pyrene) or health effects in animals similar to pesticides (e.g., halogenated biphenyls).

Because of the health-related impact of these pollutants, their control assumes an importance out of proportion to the weight emitted.

b. Mechanisms of Formation

By definition, the appearance of combustible material in the effluent of a combustion system reflects an inadequacy in the combustion efficiency. As described earlier, this indicates:

Inadequate residence time to complete combustion reactions.

And/or inadequate temperature levels to speed combustion reactions to completion.

And/or inadequate oxygen concentrations in intimate conjunction with fuel gases to allow oxidation to proceed to completion. This often is characterized by a lack of intense turbulence in the flow or an insufficient air supply.

These requirements affect both the formation and the persistence of combustible pollutants. In burning solids, the formation of hydrocarbon pyrolysis products, carbon monoxide, carbon char, and other species is an inherent part of the process. Thus, their appearance in the flue gas reflects flow bypassing, quenching (by contact with cool surfaces or admixture with cold gases), or thermal cracking (dehydrogenation of carbon-hydrogen compounds at elevated temperatures); all with a subsequent time, temperature, and composition history which prevents burnout.

It should be recognized that even the more complex ring compounds can be formed from the combustion of chemically simple fuels. Thus, polycyclic organic matter is found in the effluent from boilers burning natural gas. It is not known at this time whether such pollutants arise from, for example, the hydrogenation of soot (which has a complex six-membered ring structure) or from trace hydrocarbons in the fuel.

c. Concepts for Control

Much of the material which precedes and follows this section relates to design and operating optimization to effect complete combustion. In most real systems, the full attainment of this goal is either impossible or impractical and, thus, some unburned combustible will always be present in the effluent. Increasing awareness and concern by the regulatory authorities over the health implications of these emissions, however, further emphasizes the importance of careful design and operations.

3. Gaseous Pollutants Related to Fuel Chemistry

A third group of pollutants are generated by release and/or reaction of elements in the fuel. Common and important among these are sulfur oxides and hydrogen chloride.

a. Sulfur Oxides

Many waste streams and fossil fuels contain sulfur. The sulfur can be present in any or all of its many oxidation states from S^{-2} to S^{+6}. Of particular interest is the sulfur appearing as sulfides (organic or inorganic), free sulfur, or sulfur appearing in organic or inorganic acid forms. In each of these cases, the sulfur can be expected to appear in the flue gases as sulfur dioxide or trioxide. A small portion of the sulfur which exists as inorganic sulfates in the fuel or waste (e.g., gypsum-calcium sulfate) may be released by reduction reactions, especially in mass burning situations.

Depending upon the chemical composition (alkalinity) of the mineral ash residues, a portion of the sulfur oxides may be lost from the flue gas by gas-solid reaction. Also, some sulfur may remain with the ash. Typically, however, these losses are relatively small, and in excess of 95% of the sulfur (other than that appearing as inorganic sulfates in the fuel or waste) will be found in the flue gases from suspension fired combustors [48], and about 70% for mass burning systems [49], based on analogy with coal burning plants.

The proportioning of sulfur between the dioxide or the trioxide forms depends on the chemistry of sulfur in the fuel, the time sequence of temperature and composition of the flue gases, and on the presence or absence of catalytic ash material. Although cold-end chemical equilibrium considerations and excess oxygen concentrations favor oxidation to the trioxide, reaction rates are slow, and generally only 2 to 4% of the sulfur appears as the trioxide. Higher proportions of trioxide result from the burning of organic sulfonates, some heavy metal sulfates (which dissociate to SO_3 and an oxide), or to the "burning" of wastes such as discarded automobile batteries, which contain free sulfuric acid.

Sulfur oxides have importance as a pollutant due both to their health effects (especially in combination with respirable particulate matter) and to their corrosive effects on natural and man-made materials. Within the combustion system itself, sulfur trioxide will react with water vapor to form sulfuric acid, which has a dew point considerably above that for pure water. Indeed, to prevent serious corrosion (e.g., in the stack) from sulfuric acid, combustion system cold-end temperatures should be limited to a value safely above the sulfuric acid dew point.

b. Hydrogen Chloride

Chlorine appears in waste streams both in inorganic salts (e.g., sodium chloride) and in organic compounds. In the combustion of many industrial wastes and, importantly, in municipal solid wastes, a substantial quantity of organic matter containing chlorine may be charged to the furnace. In the combustion environment (usually containing hydrogen in considerable excess relative to the chlorine) the organic chlorine is converted, almost quantitatively, to hydrogen chloride-hydrochloric acid.

Sources of organic chlorine include the following:

Compound	Chlorine (wt %)	Uses
Polyvinyl chloride	59.0[a]	Bottles, film, furniture
Polyvinylidene chloride	73.2[a]	Film
Methylene chloride	82.6	Solvent
Chloroform	88.2	Anesthetic
DDT	50.0	Insecticide
Chlordane	59.0	Insecticide

[a] Pure resin.

The importance of hydrogen chloride emissions from combustion sources depends on the quantity in the fuel, but is usually small. Of importance to system designers, however, is the high solubility of hydrogen chloride in scrubber water (or condensed "dew" on cold-end surfaces) with well-demonstrated acid attack and chloride corrosion of metal surfaces.

4. Nitrogen Oxides

Nitric oxide (NO) is produced from its elements at the high temperatures obtained in furnaces and incinerators. At lower temperatures, NO formation is limited by both equilibrium (which favors dissociation to the elements) and kinetics. Although only a small portion of the NO further oxidizes to nitrogen dioxide (NO_2) within the furnace, oxidation does take place slowly after leaving stack at the temperatures and high oxygen concentrations of the ambient atmosphere. The air quality impact of the nitrogen oxides (referred to collectively as NO_x and reported as NO_2) arises from their participation in atmospheric chemical reactions. These reactions, especially those stimulated by solar ultraviolet light (a photochemical reaction), produce a variety of oxygenated compounds which account for the visibility reduction and eye irritation associated with smog.

In combustion systems, nitrogen oxides arise through fixation of nitrogen from the combustion air with oxygen (thermal generation). Also, NO_x is formed by oxidation of nitrogen entering the system bound in the fuel (fuel nitrogen generation). At very high temperatures, the dominant source of NO_x is thermal generation but, at lower temperatures, fuel nitrogen mechanisms dominate. The keys to the distribution among these mechanisms are the equilibrium and kinetic relationships which control the process.

a. Thermal Generation

The fixation of nitrogen with oxygen occurs by the following overall chain reaction mechanism after Zeldovitch [50].

$$O_2 + M \rightleftharpoons 2O + M \tag{55a}$$

$$O + N_2 \rightleftharpoons NO + N \tag{55b}$$

$$N + O_2 \rightleftharpoons NO + O \tag{55c}$$

and, in fuel-rich flames by the additional reaction

$$OH + N \rightleftharpoons NO + H \tag{55d}$$

The chain initiating step [Eq. (55a)] involves the collision of oxygen molecules with other molecules in the gas (represented by "M") to form oxygen radicals. The treatment of the mechanism equations [Eqs. (55a),(55b), and (55c)] for the presence of excess air, results in the following equation [51] for the <u>net</u> generation rate of NO:

$$\frac{d(NO)}{dt} = 6 \times 10^{16} T^{-1/2} \exp\left(\frac{-69,090}{T}\right) (N_2)(O_2)^{1/2} \quad \text{kg mol liter}^{-1} \text{ sec}^{-1} \tag{56}$$

where (NO), (N_2), and (O_2) are in mole fractions and T is in degrees Kelvin.

Equilibrium considerations lead to the following relationship for the overall reaction $N_2 + O_2 \rightleftharpoons 2 NO$:

$$K_p = \frac{(NO)^2}{(N_2)(O_2)} = 21.9 \exp\left(\frac{-43,400}{RT}\right) \tag{57}$$

The rate of formation of NO is significant only at temperatures in excess of 1800°C due to kinetic limitations, and doubles for every increase in flame temperature of about 40°C. Thus NO_x emissions are encouraged by high flame temperatures (e.g., with air preheat) and high excess air. NO_x can be reduced with water or steam injection or flue gas recirculation (to lower flame temperatures); by operation at low excess air (to reduce oxygen concentrations); or by staged combustion where the fuel is partially burned, heat is withdrawn through boiler surfaces, and then the rest of the required air for combustion is added (the overall effect is to lower the peak temperatures attained after the combustion gases contain a net excess of oxygen). Also, burner designs which reduce the intensity (volumetric burning rate) of combustion, produce relatively long diffusion flames, or encourage either two-stage combustion or low-temperature gas recirculation result in low NO_x emissions.

b. Fuel Nitrogen Generation

Recent studies by the U.S. Environmental Protection Agency have shown the important contribution of fuel nitrogen sources to NO_x emission from combustion systems. The mechanism by which the fuel nitrogen is converted into nitrogen oxides is imperfectly understood at present. Careful laboratory experiments, however, have typically shown that 15 to 100% of the fuel nitrogen can be converted with the higher conversion efficiencies obtained when the fuel nitrogen content is low (<0.5%) or when the combustor is operated lean [52].

A kinetic evaluation by Soete, reported by Bowman [53], which allows an estimation of the fraction Y of fuel nitrogen N_f converted to NO [i.e., Y = concentration of nitric oxide (NO) divided by the initial concentration $(N_f)_O$ of fuel nitrogen is given by

$$Y = \left[\frac{2}{1/Y - \left\{ [2.5 \times 10^3 (N_f)_O]/[T \exp(-3150/T)(O_2)] \right\}} \right] - 1 \qquad (58)$$

The demonstrated importance of fuel nitrogen in increasing NO_x emissions is of particular importance for waste incineration. Organic liquid wastes may contain solvents (e.g., pyridene, amines, or other chemicals) with a substantial "fuel nitrogen" content.

c. Emission Estimation

Because nitric oxide emissions are determined by flame temperature it has been observed [54] for a variety of fuels that the heating value of the fuel provides a better correlating parameter than the mass of fuel. Data are therefore usually converted to units of kilograms of NO_x (expressed as NO_2) per million kilocalories. The median emission rate for the small scale PHS tests [43] is 1 kg NO_x per 3.27×10^6 kcal heat release, or 0.31 kg NO_x per 10^6 kcal. Correlations of NO_x formation that have been proposed by PHS for coal, gas, and oil [54] are:

$$\text{kg } NO_x/\text{hr} = \left(\frac{\text{kcal/hr}}{3.24 \times 10^6} \right)^{1.18} \qquad \text{for gas} \qquad (59)$$

$$\text{kg NO}_{\text{x}}/\text{hr} = \left(\frac{\text{kcal/hr}}{2.18 \times 10^6}\right)^{1.18} \quad \text{for oil} \tag{60}$$

$$\text{kg NO}_{\text{x}}/\text{hr} = \left(\frac{\text{kcal/hr}}{1.87 \times 10^6}\right)^{1.18} \quad \text{for coal} \tag{61}$$

If the emission rate of 1 kg NO_x per 3.27×10^6 kcal associated with the 2.7 tonne/day PHS municipal refuse test incinerator (0.28×10^6 kcal/hr with 2470 kcal/kg refuse) and the corresponding NO_x emission rate (0.0857 kg/hr) are cast into the same functional form as Eqs. (11), (12), and (13), the resulting constant term within the brackets is 2.26×10^6. However, it must be recognized that this correlation provides only rough estimates, since factors such as furnace heat loss and excess air which are known to influence NO_x production are not taken into account. That the data for the three fossil fuels are correlated best with an exponent on firing rate greater than unity is a consequence of the increase in the mean emission factor with increases in furnace size.

The PHS [54] obtained a comparable correlation for refuse combustion in small units with an exponent of 1.14 in place of 1.18. In the absence of sufficient data to obtain a reliable estimate of the exponent for municipal-scale refuse incinerators, the value of 1.18 is recommended to extrapolate data from one size to another.

Emission factors for the combustion of wastes other than municipal refuse can be approximated by analogy with the combustion processes for gas, oil, coal, and (mass burning) refuse. If waste analysis indicates a fuel nitrogen content greater than, say, 0.2%, and/or operation will be at a relatively high excess air level (as is often the case for waste combustion), consideration should be given to increasing the estimate above the emission rate, calculated using Eqs. (59), (60), or (61). An estimate of the maximum NO_x emission can be made using the equilibrium relationship [Eq. (57)], evaluated at the adiabatic flame temperature. Experience shows, however, that such a technique produces unrealistically high estimates.

Chapter 4

COMBUSTION SYSTEMS

Waste incineration typically occurs as a hot turbulent flow within a refractory lined or water cooled (boiler) enclosure. This chapter provides a framework of basic understanding of the materials, heat transfer, and mixing processes which should ideally be incorporated into the design.

A. ENCLOSURES

The enclosure, the furnace itself, plays a key role in assuring adequate performance of a thermal processing system. Specifically, the enclosure and its characteristics affect the process in the following ways:

Shape. Enclosure shape affects the radiative flux onto incoming fuel (affecting ignition time and flame length), the gross flow patterns (including recirculation and bulk mixing), and the heat absorption patterns in boiler type enclosures.

Volume. Enclosure volume determines the mean residence time, affecting burnout of CO, H_2, soot, tars, and other combustible pollutants.

Accessory Features. The enclosure may include heat absorbing surfaces for energy recovery; air jets for overfire combustion and temperature control in mass burning systems or secondary and tertiary combustion air jets for liquid pulverized solid or gas fired units; means to feed waste and withdraw slag or solid residues; or water sprays to cool the combustion gases as a precursor to temperature sensitive air pollution control devices.

The design of incinerator enclosures is not yet a rigorous engineering discipline, especially for mass burning systems. In the geometrically simple case of axially fired cylindrical chambers, a number of analytical tools are available to describe the flow patterns (see Section C). Perhaps the greater challenge to the analyst is to <u>simplify</u> the problem to one susceptible to analysis. One should take some comfort in the fact that hundreds of millions of dollars in incinerator equipment now operating is based on little more combustor design analysis than the postulation of a mean residence time goal (usually 2-4 sec for gases, 30-60 min for solids) and an assembly of physical constraints brought on by commercially available components. On the other hand, many of those existing plants are only marginally operable and, as environmental controls become more stringent, will be shut down.

1. Refractory Enclosure Systems

Refractories are defined as non-metallic materials suitable for the construction or lining of furnaces operated at high temperatures. Stability (physical and chemical) at high temperatures is the primary material requirement, as the refractory system may be called upon, while hot, to withstand compressive and (limited) tensile stresses from the weight of the furnace or its contents, thermal shock resulting from heating or cooling cycles, mechanical wear from the movement of furnace contents (or even the furnace structures themselves due to thermal expansion or, in a kiln, for the entire furnace), and chemical attack by heated solids, liquids, gases, or fumes.

In most incineration systems, refractory materials constitute an important part of the furnace enclosure, ducting, and/or stack. The functions of refractory materials include:

Containment of the combustion process and flue gases in an enclosure resistant to failure from thermal stresses or degradation from high temperature abrasion, corrosion, and erosion

Reradiation of heat to accelerate drying, ignition, and combustion of incoming feed

Protection of plant personnel from burn injury through an insulating effect

Enhancement of combustion by use of refractory baffeling systems which increase turbulence in high temperature flue gas passages

Support of the burning mass or residues (hearth burning systems)

Protection of vulnerable system components in unusually severe environments (e.g., incinerator boiler superheater surfaces)

To serve these many functions, a wide range of refractory materials have been developed. Art and science must be combined for proper application of refractory systems, particularly for incineration systems where frequent and sometimes unanticipated changes in waste character can result in wide swings in temperature and ash (or slag) and gas composition.

The data and discussions presented here (drawn especially from Ref. 55) must then be recognized as only a brief introduction to the field of refractory selection and design. In the course of system design, the engineer is encouraged to consult with firms and individuals (including owners and operators of existing facilities) with experience closely paralleling the intended application. As an indication of the difficulty in evaluating refractories, it is noteworthy that refractory life or maintenance requirements for a given plant cannot be reliably forecast from pilot scale experience; even when similar wastes or fuels, system geometry, etc. are used. This simulation problem reflects the complexity of the processes which induce refractory failure.

Refractory materials are supplied mainly in:

Preformed shapes, including standard sizes of brick, most commonly 22.9 × 11.5 × 6.4 cm (9 × 4.5 × 2.5 inch) and 22.9 × 11.5 × 7.6 cm (9 × 4.5 × 3 in.) "straights"; arch, wedge, kiln block, and other shapes used to build arches or to line cylindrical chambers and flues; special and often proprietary designs incorporating hangers for suspended wall and roof construction; and special application shapes such as ignition tiles or checkerbrick

Plastic refractories (premixed and ready to use), and dry, ramming mix (to be mixed with water before using) which can be rammed into place to form monolithic structures

Castable and gunning mixes which are poured or sprayed (gunned), respectively, to form large monolithic structures

Mortar and high temperature cements

Granular materials such as dead-burned dolomite, dead-burned magnesite and ground quartzite

a. Composition of Refractories

Table 20 indicates the primary constituents of many refractories commonly used in incineration applications. It can be seen that most fireclay, alumina, and silica refractory materials are composed of mixtures of alumina (Al_2O_3) and silica (SiO_2) with minor constituents, including titania (TiO_2), magnesite (MgO), lime (CaO), and other oxides.

TABLE 20 (PART 1)

Properties of Selected Refractory Brick[a]

Type	Class	Primary constituents (wt %)			Pyrometric[b] cone equivalent	Density (kg/m³)
		Silica (SiO₂)	Alumina (Al₂O₃)	Titania (TiO₂)		
Fire clay[c]	Superduty	40–56	40–44	1–3	33–34	2240–2320
Fire clay[e]	1. High duty	51–61	40–44	1–3	33–33.5	1920–2325
	2. Medium duty	57–50	25–38	1–2	29–31	1920–2325
	3. Low duty	60–70	22–33	1–2	15–29	1920–2325
Fire clay[f]	Semi–silica	72–80	18–26	1–2	29–31	1840–2000
High alumina[g]	1. 45–48%	44–51	45–48	2–3	35	2240–2565
	3. 60%	31–37	58–62	2–3	36–37	2485–2885
	5. 80%	11–15	78–82	3–4	39	2645–3045
	7. 90%	8–9	89–91	0.4–1	40–41	2325–2645
	8. Mullite	18–34	60–78	0.5–3	38	2325–2645
	9. Corundum	0.2–1	98–99+	Trace	42	2725–3205
Silica[h]	1. Superduty	95–97	0.1–0.3[i]		N/A*	1680–1890
	2. Conventional	94–97	0.4–1.4[i]		N/A*	1680–1890
	3. Lightweight	94–97	0.4–1.4[i]		N/A*	960
Silicon carbide[j]	1. Bonded	Silicon carbide			38	2325–2645
	2. Recrystallized				N/A*	—
Insulating[k]	1. 870°C	Variable:			N/A*	580
	3. 1260°C	alumina,			N/A*	740
	5. 1540°C	fireclay,			N/A*	940
	6. 1650°C	perlite, etc.			N/A*	960
Chrome[l]	1. Fired	Cr₂O₃ (28–38), MgO (14–19)			N/A*	2960–3285
	2. Chemically bonded	Al₂O₃ (15–34), Fe₂O₃ (11–17)			N/A*	—

TABLE 20 (PART 2)

Type	Modulus of rupture (kg/m²)	Cold crushing strength (kg/m²)	Thermal conductivity kcal hr⁻¹ m⁻² (°C/cm)⁻¹			Thermal expansion (% at 1000°C)
			@500°C	@1000°C	@1500°C	
Fire clay[c]	2930–7900	7320–29,300	1.16	1.25	N/A[d]	0.6
Fire clay[e]	2440–14,650	7320–34,200	1.11	1.22	N/A*	0.6
	3900–12,200	8300–29,300	N/A	N/A	N/A	0.5–0.6
	4880–12,200	9760–29,300	N/A	N/A	N/A	0.5–0.6
Fire clay[f]	1460–4400	4880–14,650	0.89	1.09	N/A*	0.7
High alumina[g]	4880–7800	12,200–29,300	–	–	–	–
	2930–8790	8790–34,200	1.20	1.25	1.29	0.6
	5370–14,650	19,530–43,900	–	–	–	0.7
	5860–17,100	19,520–43,900	1.92	1.65	1.65	0.7
	4880–17,100	17,100–43,900	–	–	–	–
	8790–14,650	24,400–43,900	3.40	2.25	2.32	0.8
Silica[h]	2440–4880	7320–17,100	1.19	1.55	1.90	1.3
	2930–5860	8790–19,530	1.00	1.25	1.61	1.2
	N/A	N/A	0.40	0.65	1.00	–
Silicon carbide[j]	9760–19,530	12,200–73,200	15.61	12.22	8.47	0.5
	N/A	N/A	–	–	–	–
Insulating[k]	340–490	440–540	0.137	0.275	N/A*	Low
	490–830	540–930	0.251	0.338	N/A*	Low
	830–1465	830–1470	0.312	0.395	0.482	Low
	1950–2930	3900–4880	0.344	0.426	0.512	Low
Chrome[l]	3420–6350	9760–19,530	2.04	1.83	1.75	0.8
	3420–8790	9760–24,400	–	–	–	–

FOOTNOTES TO TABLE 20

[a] Source: selected manufacturers data.

[b] See Appendix C.

[c] These materials exhibit the highest refractoriness of any fireclay brick; stable volume; good spall resistance; good load bearing properties; fair resistance to acid slags, but only moderate resistance to basic slags.

[d] N/A = not available; N/A* = not applicable or beyond meaningful range.

[e] High duty has low thermal expansion, fair resistance to acid slags and spalling, lower resistance to basic slags. Properties degrade for lower duty brick.

[f] These materials exhibit rigidity under load, volume stability, and high resistance to structural spalling, and to attack by volatile alkalis.

[g] These materials exhibit high refractoriness increasing with alumina content; high mechanical strength; fair to excellent spall resistance; high resistance to basic slags, and fair resistance to acid slags.

[h] These materials exhibit high refractoriness, mechanical strength, and abrasion resistance; and low thermal spalling above 650°C. "Hot patch" brick has good spall resistance. This brick is resistant to acid slag but is attacked by basic slag.

[i] Remainder primarily lime.

[j] These materials exhibit high refractoriness, thermal conductivity, resistance to spalling, slagging, and abrasion. They oxidize at critical temperatures.

[k] These materials exhibit high resistance to thermal spalling but are degraded rapidly by abrasion or slag. Most often used as a backing for more resistant bricks.

[l] These materials exhibit high resistance to corrosion by basic and moderately acid slags. Iron oxide can cause spalling.

The minerals which constitute refractory materials exhibit properties of acids or bases as characterized by electron donor or receptor behavior. Acidic oxides include silica, alumina, and titania and basic oxides include those of iron, calcium, magnesium, sodium, potassium, and chromium. The fusion temperatures and compatibility of refractory pairs or slags and refractory is related especially to the ratio of the weight percentages of basic to acidic constituents.

The mineral deposits from which the fireclays are obtained include (1) the hydrated aluminum silicates (e.g., kaolinite) and flint clays which contribute strength, refractoriness, and dimensional stability, and (2) the plastic and semiplastic "soft clays" which contribute handing strength. Bauxite or diaspore clays, precalcined to avoid high shrinkage in the firing of refractory products, are the principle source of alumina for high alumina refractory (greater than 45% alumina).

Basic refractories incorporate mixtures of magnesite (usually derived from magnesium hydroxide precipitated from sea water, bitterns, or inland brines), chrome ore, olivine [a mineral incorporating mixed magnesium, silicon, and iron oxides with forsterite (2 $MgO \cdot SiO_2$) predominating], and dolomite, a double carbonate of calcium and magnesium.

b. Properties of Refractories

(1) Refractory Structure

At room temperature refractory products consist of crystalline material particles bonded by glass and/or smaller crystalline mineral particles. As the temperature increases, liquid phase regions develop. The mechanical properties of the refractory at a given temperature depend on the proportion and composition of the solid minerals, glassy structure, and fluid regions.

In the manufacture of refractory products, unfired or "green" refractory masses are often heated or "fired" in a temperature controlled kiln to develop a ceramic bond between the larger particles and the more or less noncrystalline or vitrified "groundmass." In firing, a high degree of permanent mechanical strength is developed. For any given refractory composition, however, there exists an upper temperature limit above which rapid and sometimes permanent changes in strength, density, porosity, etc. can be expected. This upper limit is often a critical parameter in the selection of refractory for a given service.

The evaluation of the high temperature softening behavior of fireclay and some high alumina refractory materials is often accomplished by determination of the Pyrometric Cone Equivalent or PCE (see Table 20 and Appendix C). In this test, a ground sample of material is molded into a test cone, mounted on a ceramic plate tilted to a slight angle, and heated at a definite rate. PCE standards mounted adjacent to the test cone are observed. The PCE of the material being tested is defined as that of the standard cone whose tip touches the plaque at the same time as that of the test specimen. Since softening of refractory is affected by the chemical nature of the surrounding atmosphere, the test is always carried out in an oxidizing atmosphere. If the situation under design is reducing in nature, consideration should be given to evaluation under similar atmospheric conditions (e.g., see Table 42, Chap. 5).

(2) Melting Behavior

Alumina-silica refractories (e.g., fireclay and high alumina brick) contain both crystalline and glassy material. As temperatures exceed approximately 980 to 1095°C, the glassy bond begins to become progressively less rigid and, indeed, becomes a viscous liquid. As viscosity drops, the glass lubricates relative motion of the crystalline particles and, under stress, deformation can occur. At higher temperatures, the crystalline material begins to dissolve in the glass and, finally, the mass is unable to retain its original shape, deforming under its own weight.

The temperature level at which deformation becomes critical to structural integrity depends particularly on the stress level, the availability of cooler regions within the refractory shape which, since more rigid, will accept the load, and on the alumina content (since the crystalline alumina granules are the last to melt). Also, a past history of long-term, high temperature soaking may have resulted in the solution of solid material into the glassy phase with a consequent increase in the observed viscosity at a given temperature. This phenomena may explain in part why a hard-burned fireclay is generally more resistant to deformation than a light-burned or unused brick.

Silica refractories which are almost wholly crystalline when cold differ greatly from fireclay-based refractories in their melting behavior. In silica refractories, the melting is due especially to the fluxing action of alumina, titania, and alkali. Indeed, if the flux content is less than 0.5%, the formation of a substantial (>10%) liquid fraction (and associated deformation and strength losses) occurs abruptly over a 50°C temperature range near 1700°C.

The melting characteristics of basic refractories are largely determined by the melting of the groundmass surrounding the highly refractory magnesite, chrome spinel, and forsterite granules. Thus, this class of material exhibits a wide range of refractoriness showing, for example, failure temperatures in the ASTM Method C-16 Load Test from 1290 to 1705°C.

(3) Dimensional Changes

Almost all green refractory materials exhibit permanent dimensional changes on initial heating. As firing proceeds, the rate of change of dimensions diminishes. If actual use reflects a careful choice of refractory (and, naturally, a specific curing cycle), little significant dimensional change will be observed in service.

As with most materials, refractories also exhibit reversible expansion and contraction when heated and cooled, respectively. An indication of the linear expansion at elevated temperatures for refractory materials is given in Table 20.

(4) Abrasion and Impact

In the incineration of solids, refractories are often subject to the impact of heavy pieces of material charged into the furnace, abrasion by moving solids, or direct impingement by abrasive fly ash suspended in fast moving gas streams. Abrasion and impact resistance are often evaluated using standard rattler tests (ASTM C-7) and, more recently, by blasting the surface using silica or silicon carbide grit. Performance in the latter test has been shown to correlate well with modulus of rupture data.

(5) Spalling

Excessive thermal or mechanical stresses which are applied near to the face of a refractory brick or monolithic structure or adverse changes in the internal structure of the refractory in zones paralleling the hot face can lead to the loss of fragments (or spalls) of surface material. If the newly exposed refractory material is also overstressed or undergoes a second cycle of internal structural change, the degradation process will continue until structural failure occurs or until heat losses through the thinned brick require shutdown. The loss of face refractory through cracking and rupture is known as spalling. Spalling is of three general types (thermal, mechanical, and structural).

(a) Thermal Spalling Thermal spalling results from stresses generated within the refractory arising from the unequal extent of thermal expansion (or contraction) due to temperature gradients. Failure of the refractory is usually associated with rapid changes in temperature. Thus, thermal spalling failures can be traced to material and/or operational factors.

Refractory with low thermal expansion properties, high tensile strength, and a maximum extensibility (the ratio of tensile strength to modulus of elasticity) show superior thermal spalling resistance.

For fireclay brick, spalling characteristics are largely dependent upon the proportion of free silica in the constituent clays; the size and size distribution of the particles; and upon the amount and composition (softening temperature) of the glassy bond. Up to 980 to 1095°C (1800 to 2000°F) for example, a high glass content will result in a high spalling tendency. Above these temperatures, the glassy materials soften (allowing stress relief), and the fireclay brick becomes very resistant to spalling.

Since the extent and composition of the glassy bond is affected by the time and temperature history of the brick during its manufacture, and by the temperatures experienced in service, consideration must be given to both factors in evaluating the potential for spall failure. Table 21 indicates some general trends.

The specific temperature (650°C) where spall sensitivity is abruptly improved for silica brick reflects the abrupt volume change accompanying the crystalline inversion of the cristobalite. Above 650°C spalling resistance is excellent.

Among the basic brick compositions, the best spall resistance is found in brick with a high periclase content: bonded with magnesium aluminate spinel or a chrome spinel. In some applications, the basic brick is encased

TABLE 21

Thermal Spall Resistance of Refractory Materials

Type of brick	Comparative spall resistance
Fire clay:	
Superduty:	
Spall resistant	Superior
Regular	Excellent
High Duty:	
Spall resistant	Good
Regular	Fair
Semi-silica	Good
High alumina:	
45 to 60% alumina	Superior to good
>60% alumina	Superior to fair
Silica:	
All types	Excellent above 650°C
General duty	Poor below 650°C
Spall-resistant	Fair below 650°C
Basic:	
Chrome, fired	Fair
Chrome-magnesite (chemically bonded or fired)	Good to fair
Magnesite, fired or chemically bonded	Fair
Magnesite, fired (high periclase)	Excellent
Magnesite-chrome (chemically bonded)	Superior to excellent
Magnesite-chrome, fired	Very superior to excellent
Spinel bonded magnesite, fired	Superior
Forsterite	Fair

[a]Source: Ref. 55.
[b]Spall resistance ratings are comparative <u>only</u> within groups.

in a metal sheath. At low temperatures, the case holds the fragments in place. At higher temperatures, the metal oxidizes, reacts with the magnesia and bonds adjacent brick together.

(b) <u>Mechanical Spalling</u> Rapid drying of wet brickwork; inadequate provision for thermal expansion; or pinching of hot ends (especially in arches) can lead to stress concentration and failure. "Pinch spalling" is common on the inner surfaces of sprung arches since the hot faces expand more than the cold ends.

(c) <u>Structural Spalling</u> Structural spalling refers to losses of surface material due to changes (in service) in the texture and composition of the hotter portions of the brick. These changes can be brought about through the action of heat; through reactions with slags or fluxes; or through the reactions of or with gases which permeate the brick [see Section A.1.a.(6d)]. The changes resulting in spalling are those where zoned structures are formed which show markedly different thermal expansion characteristics, increased sensitivity to rapid temperature change, or the presence of shrinkage or expansion cracks.

   (6) Chemical Reactions

   At the temperatures found in incinerators and many other combustion systems, chemical action can be one of the greatest factors contributing to the ultimate destruction of refractory structures. The mineral constituents of wastes (or fuels) or the gases in the furnace enclosure may penetrate the refractories; changing the size and orientation of crystals, forming new minerals or glass, altering the texture and physical or chemical properties of the brick, or dissolving (corroding) granules or bonding structures. The chemical agents which participate in these chemical reactions are known as <u>fluxes</u> and the corrosive melting action which ensues as fluxing.

   The chemistry of the mineral residues from the incineration of wastes and the products derived from reactions between the residues and refractory materials are exceedingly complex. Indeed, the chemical, physical, and temporal parameters affecting corrosion in refractories is so complex that it has not been found possible to adopt a single screening test of general utility. Experience however gives insight into general principles of refractory-residue compatibility.

(a) <u>Acidic and Basic Fluxes.</u> In evaluating the potential for refractory reactions, the most useful general concept matches acidic slags with acidic refractories and basic slags with basic refractories. In this context, the oxides of the following elements show acidic or basic properties:

Acidic oxides	Basic oxides
Silicon (as SiO_2)	Iron (as FeO)
Aluminum (as Al_2O_3)	Calcium (as CaO)
Titanium (as TiO_2)	Magnesium (as MgO)
	Potassium (as K_2O)
	Sodium (as Na_2O)
	Chromium (as Cr_2O_3)

Since fluxing involves chemical reactions, the analyst should take care to evaluate the acid-base balance using composition expressed in mole percent instead of the commonly reported weight percent.

Refractory brick falls into acidic or basic categories as follows:

Acidic brick	Basic brick
Fireclay	Chrome
High-alumina	Magnesite
Silica	Forsterite

The use of the net acidic or basic chemistry of a slag is a good but not a perfect criterion for refractory selection. Exceptions are common due to differences in operating and reaction temperatures, reaction rates, viscosities of reaction products, and the formation of protective glazes and coatings on the refractories. For example, the relative concentrations of the elements in a slag could at, say, low concentrations yield a reaction product of high melting point while at high concentrations, a different reaction product would be formed with a very low melting point. This example is found in practice in the pattern of attack of silica refractories by ferrous oxide slags.

(b) Corrosion Rate. The corrosion rate of refractories depends on a number of factors, factors affecting the rate at which reactants are brought together, react, and are carried away.

The rate at which reactants are brought together is inhibited by (1) low concentrations of fluxes, (2) the diffusional resistance of protective layers or dense brick structure, (3) the adherence or high viscosity of reaction products, (4) the inability of the flux to "wet" the surface of the refractory, (5) temperatures low enough so that the flux is in the solid state, and

(6) lack of agitation or turbulence in the molten fluxing agent (usually induced by impinging or shearing gas flows).

The rate of reaction, as with other chemical reactions, is strongly
dependent on temperature. In some cases, only a few degrees rise in
temperature will cause destruction of an otherwise satisfactory lining.
This sensitivity emphasizes the importance of making prompt and thorough
evaluation of refractory response to changes in the operation of combustion
systems which increase average temperatures or which increase the thermal loading in "hot spot" regions (e.g., at flame impingement points).

The rate at which reaction products flow away (thus exposing fresh
surface to attack) is reduced by (1) high viscosity reaction products (note
that molten minerals increase in viscosity as the temperature falls so that
wall cooling, obtained with natural or forced convection or imbedded water
filled cooling coils, will freeze the slag and impede corrosion). If this
approach is used, however, consideration must be given to thermal spalling
when the furnace is cooled off, or to mechanical spalling if slag accumulations become heavy; (2) special selection of brick composition, texture,
and/or permeability to avoid spalling due to the increased volume of reaction products. Such structural spalling can occur upon the reaction of
chrome bearing basic brick with Fe_2O_3 or the reaction of soda or potash
fumes with fireclay and some high alumina brick.

(c) Reactions Between Refractories. Just as "foreign materials" such as
fly ash can react with refractories, so refractory bricks and/or mortars of
different compositions can enter into reactions with one another. The potential for reaction depends on physical contact; the existence of low-melting
eutectics such as the alumina-silica eutectic with a melting point of 1595°C
(2903°F); the impurities within either refractory which can form an inter-
stitial liquid phase even though the brick as a whole appears strong and
rigid; by the availability of mechanisms such as gravity, diffusion, or capillary action to mingle the brick components; and by the temperature level at
the interface.

Engineering guidelines for the abutting of different refractories and
for the choice of mortars at the interface are best developed with refractory manufacturers in consideration of the overall system requirements.

(d) Reactions with Gases. Furnace gases can have a substantial impact
on refractory-slag and refractory-refractory reactions; affect the melting
range and viscosity-temperature properties of slags; and can, themselves,
participate in refractory attack. From the sections of this book dealing
with combustion system fluid flow, it should be recalled that the composition
of furnace gases is seldom constant at any one point in the system and cer-
tainly varies from point to point in the system. Thus, if refractory attack
by gases can be a factor in a given design, the refractory selection will

include consideration of the range of furnace atmospheres to be encountered
and not only the "average."

The most important measure of furnace atmosphere corrosive impacts
relates to its oxidizing or reducing nature. For most refractory oxidizing
atmospheres are preferred. An exception is silicon carbide which slowly
oxidizes above 900°C (1650°F). Oxidizing agents for silicon carbide include
oxygen, carbon dioxide, and water vapor. To retard these oxidation reactions, the silicon carbide grains are usually coated with ceramic material.

Reducing atmospheres tend to enhance the corrosivity of iron bearing
slags, perhaps due to the greater basicity of FeO versus the more fully
oxidized forms. The relative ease with which iron oxides are reduced and
oxidized and the resulting changes in chemical reactivity and density of the
oxides accompanying these changes can weaken brick exposed to alternating
oxidizing and reducing atmospheres. Also, between 370 and 540°C (700 to
1000°F) carbon monoxide can decompose (catalyzed by iron oxides) into
carbon and carbon dioxide. Iron oxides may also be partially reduced to
metallic iron or iron carbide. The consequent volume changes can lead to
rapid disintegration of the brick.

Some superduty and high-duty fireclay brick and high-alumina brick,
if fired to Cone 18 or higher under controlled conditions, offer high resistance
to carbon monoxide effects, apparently due to the conversion of iron oxides
into noncatalytically active silicates.

Unburned or partially cracked natural gas can undergo decomposition
and carbon deposition within refractories in the temperature range of 480
to 820°C (900 to 1500°F).

Chlorine reacts with some components of various refractories (e.g.,
alumina) at about 950°C (1750°F) to form volatile chlorides. Alumina also
reacts with sulfur oxides (in the presence of water vapor) in the cooler
parts of the linings to form aluminum sulfate. The volume increase accompanying the "sulfating" reaction will weaken the brickwork.

Clearly, the permeability to gases of refractory brick will provide a
useful measure of the potential extent and importance of reactions with furnace gases. While permeability is related to porosity, the relationship is
not direct since the orientation and diameters of pores and the extent to
which they interconnect affect gas diffusion rates. Operating history which
could include sealing of the surface with a relatively continuous glaze or
the loss of binders (such as occurs with chemically bonded basic brick)
also influences gas permeability. Thus, experience is often the only means
to assess performance.

c. Selection of Refractories

In most furnaces and particularly in incinerators, it is of value to give
careful attention to the selection of "optimal" refractory. Optimal is in

quotation marks to highlight the definition of optimal as not universal but reflecting the special needs and values of each user. Thus, total cost per unit of production (taking into account initial materials and installation costs, interim maintenance costs, and the value of lost production during refractory related maintenance or rebricking) may be a prime consideration. For waste disposal systems, reliability and availability of the incineration system has value which far exceeds the differential material costs between inferior and superior refractory grades, especially since both transportation and installation expense is essentially equal for each grade.

Beyond economic factors, refractory selection should involve a detailed analysis of operational, design, and construction facts and expectations. Ideally, the designer should carefully analyze the chemical, fluid flow, and heat flow parameters which characterize the proposed system and then interact with specialists in refractory design and selection. Often, parallel experience will be available to guide the selection process. Also, it is usually found that only one or two among the several refractory properties are keys to success (e.g., refractoriness alone, resistance to thermal shock, or slagging). As mentioned in the introduction to this discussion and restated here for emphasis, refractory selection is an expert's game, derived especially from broad, in-depth experience and only just now emerging as a quantitative technology.

2. Water Cooled Enclosure Systems

Containment of combustion processes in refractory lined chambers limits the peak temperatures which can be permitted (requiring increased excess air levels and, consequently, larger air pollution control and fan equipment). An alternative enclosure concept uses metal surfaces protected against overheating by water cooling.

a. External Cooling

In its simplest embodiment, cooling of the metal furnace walls is accomplished by flowing water, uncontained, over the outside surface. For example, a small, rotary kiln-type furnace with coarse water sprays playing on the shell has been successfully used for industrial waste incineration. Scale, the deposit of minerals in the make-up water, builds up over time and spontaneously (or with "help") cracks off.

Systems using such a cooling technique are limited in size and heat release rate because of uneven heat transfer in areas where adherant scale is formed. In such "insulated" areas, wall temperatures increase rapidly and metal wastage occurs. The low capital cost and ease of repair of these furnaces, however, makes them attractive in special cases.

b. Boiler Enclosures

In larger combustors or when heat recovery is profitable, the cooling water, under pressure, is passed into pipes or tubes immersed in the hot gas stream or arranged in panels lining the furnace wall. Often, the wall tubes are welded together with a narrow steel strip between the individual tubes to form a continuous, gas-tight membrane or <u>waterwall</u> enclosure.

To better understand the system, let us follow the course of the water entering the boiler plant.

(1) Water Treatment

Raw water, containing dissolved minerals and suspended matter would be an unsatisfactory feed to a boiler. To avoid scale build-up and/or corrosion from these contaminants, the water must be treated with the level of treatment increasing with the severity of the water side environment (as characterized by the temperature and pressure of the product steam). Treatment methods include filtration, softening, distillation, and/or ion exchange.

(2) De-aeration

Water also contains dissolved gases (air components, CO_2, etc.) which would accumulate in the boiler after a time and which increase the minimum pressure obtainable when the water is condensed. (In steam turbines operated with a condenser, the steam can, theoretically, be expanded to the pressure corresponding to the vapor pressure of water at the condenser temperature. Noncondensable or "fixed" gases accumulate and pressurize this area and must be pumped out. Their net effect is to reduce the efficiency of the turbine.)

To remove most of the dissolved gases, the treated water is heated with steam or electricity to the atmospheric boiling point in a deaerator. The water leaving the deaerator is ready for introduction into the boiler using the feedwater pumps to raise the pressure to the boiler's working level. In some steam plants, low pressure "waste steam" is condensed to preheat the boiler feedwater after the deaerator but prior to entering the boiler proper.

(3) Boiler

At the point of introduction into the boiler, the feedwater is treated, deaerated, and, perhaps, somewhat preheated such that its temperature is in the range 100 to 200°C. In passing the water through the boiler, it is desirable to optimize the temperature difference between the water and the hot combustion gases (maximum heat transfer rate) to minimize the required amount of heat transfer area (capital cost) while still extracting the maximum amount of heat from the combustion gases. In larger boilers, this will include such components as:

Economizer: one or more banks of tubes between which the hot flue gases flow and connectively transfer heat to the feedwater. The feedwater is usually not heated, however, to the point where evaporation occurs. The economizer is located in the part of the boiler where the flue gas temperature is the lowest.

Convection Boiler: one or more banks of tubes ("or passes") between which the hot flue gases flow where the water (from the economizer) is evaporated. Flow through the tubes may be due to buoyancy effects (natural convection) or pumps (forced convection) and is two-phase: containing both liquid water and steam.

Radiant Boiler: the water walls and, in some cases, banks of tubes exposed to the combustion zone between which the hot flue gases flow. Heat transfer rates are very high and radiant energy transport from the incandescent refuse bed, flame, and/or hot gases (see subsequent sections of this chapter) is the predominant means of heat transfer.

Steam Drum: one or more large accumulators with disengagement space and mechanical devices to separate the gaseous steam from the liquid water. The latter is recirculated to the convection or radiant boiler sections. The product steam, in thermodynamic equilibrium with liquid water, is "saturated" at the temperature and pressure of the steam drum contents.

Superheater: a radiantly and/or convectively heated tube bank where, at (roughly) constant pressure, the steam is further heated to produce dry steam with a heat content greater than that of the saturated steam. Such superheated steam conditions are often advantageous as they will produce more mechanical energy in a turbine than will saturated steam and/or will tolerate moderate heat losses without condensation in, say, a steam distribution system pipeline.

(4) Metal Wastage

In boilers using wastes as fuels (and for conventional liquid or solid fuels as well), metal wastage due to corrosion and erosion and tube fouling due to the buildup of deposits have presented serious problems to the system designer and operator. Detailing the nature and cures for such problems is beyond the scope of this book and, at this writing, is still a matter of intense study and speculation. Several basic concepts, however, merit qualitative description:

Low Temperature Corrosion In regions of the boiler, especially in the economizer or on any surfaces used to transfer heat to incoming combustion air (the air heater), slow-flow zones (low heat transfer rates), and/or too cold feed water or air can lower metal temperatures to the point where condensation of moisture in the flue gases will occur. The presence of mineral acids (e.g., sulfuric or hydrochloric acid) in the flue gas (especially sulfur trioxide) leads to condensation at temperatures considerably above 100°C. The resulting metal wastage rate, accelerated by the presence of soluble chlorides or acids, can become unacceptably high. Clearly, such corrosive mechanisms are always operative during boiler start-up and shut-down.

The "cure" for this type of corrosion is straightforward: design to avoid dead zones and to maintain metal temperatures safely above the dew point of the flue gases. Also, minimize the frequency and duration of cooldowns.

High Temperature Corrosion   In regions of the furnace above, say, 400°C, a variety of mechanisms for chemical attack of the metal tube surfaces can become operative.

Chlorine and Sulfur Attack   Chlorine, appearing in the flue gases as hydrochloric acid (e.g., from the combustion of chlorinated hydrocarbon wastes or polyvinyl chloride) or in salts such as sodium or potassium chloride has been shown to participate in corrosive attack of metal tubes. Sulfur, appearing as the dioxide or trioxide or as sulfates appears to slow the rate of attack of the metal by chlorides [59]. While the exact mechanism of attack is still in question, it is clear that fireside metal temperature is the single most important parameter in evaluating the potential for rapid metal wastage. The flue gas temperature, however, is a second, though less important, variable at gas temperatures of interest.

Data reported by Battelle [56-61] indicated that a maximum fireside metal temperature of 205°C (400°F) should give long (say, 15 years) carbon steel boiler tube service for systems burning 100% municipal refuse. Allowing for a 25°C temperature drop across the tube wall (a reasonable average for boiler tubes containing liquid water and thus experiencing a high heat transfer rate on the inner wall, but far too little for tubes containing gaseous steam) this corresponds to a maximum (saturated) steam pressure of 9.85 atm (130 psig). For higher pressures and temperatures, increased wastage must be accepted and/or more costly tube metal alloys must be used.

The problems of metal wastage are of special concern for superheater surfaces. The lower heat transfer rates of the steam (compared to liquid water) results in higher fireside tube temperatures and greatly accelerated corrosion. Some relief from this problem has been found by the use of rammed silicon carbide type refractory coatings on the tubes [62]. This "solution" comes at the cost of lowered heat transfer rates and increased investment and maintenance expense.

Oxidizing and Reducing Conditions   In the incineration of highly non-ideal fuels such as raw municipal refuse and especially for mass burning configurations, it is common to produce flue gases which fluctuate in composition between oxidizing (having an excess of oxygen) and reducing (absence of oxygen and the showing presence of carbon monoxide, hydrogen, hydrocarbon gases, etc.). Metal surfaces exposed to such changing gas compositions are subject to rapid wastage. In large part, this wastage is due to the effects of repeated cycles of surface metal oxide formation and then reduction. Flaking of the weak reduced metal structure is accelerated by the "shot blasting" effect of entrained particulate.

In mass burning systems, the bed processes always produce reducing gases and, consequently, sidewalls and radiant tube banks are particularly prone to this type of attack. Protection of the sidewalls with refractory up to a distance 3 to 10 m above the grate line has been used success` .y to cure the sidewall corrosion problem. Introduction of sufficient secondary air above the fire and stimulation of high levels of turbulence can greatly assist the burnout of reducing gases prior to their entry of the tube banks.

Abrasion (Erosion) Wastage   The mechanical erosion of tube surfaces by fast-moving fly ash particles can rapidly lead to tube failure. The fly ash from municipal refuse combustion has been shown to be particuarly abrasive (more so than, for example, most coal ash). This problem can be mitigated by reducing the velocity of the flue gases between the tubes (say, to 3.5-4.5 m/sec) by coating the tubes with refractory (or letting slag build up to a degree), and by careful design of tube bank geometry and flow patterns. In general, these remedies lead to larger, more costly boiler facilities.

Slagging   Slag build-up is not directly responsible for tube wastage (and, as mentioned above, can even reduce erosive metal losses). Slag accumulation will, however, reduce heat transfer rates and increase the pressure drop and gas velocity (erosion rate) through the boiler flues. It should be recognized that important corrosion reactions have been shown to occur within the slag layer so that slag build-up cannot be used to infer a lack of chemical attack.

Slagging can be avoided by designs which maintain fireside metal temperatures below the range where the slag becomes tacky. For municipal refuse, this range is approximately 600 to 700°C. Alternatively, the boiler passes can be equipped with "soot blowers" which use steam or compressed air jets or metal shot to periodically dislodge adherant slag. Note, however, that cleaning the tube surface can also result in the removal of coatings which were performing a protective role with respect to tube attack. Thus, following soot blower activation, corrosive wastage will be initiated, typically at very high rates, until a protective coating of slag and/or corrosion products is reestablished.

B. HEAT TRANSFER

It is beyond the scope of this chapter to deal with the complex problems of heat transfer analysis for combustors. Texts such as Radiative Transfer [27] and others are more appropriate resources. However, several heat transfer related considerations should be noted.

1. Conduction

The primary areas of application of conduction heat transfer analysis in the design of incineration systems are in estimating fuel heating rates (see Chap. 3, Section C.1.a) and in predicting heat losses and outside wall temperatures for the combustion chamber. Although the heat flux in combustion systems is largely radiatve (except the convective heating of boiler tubes), heat loss by conduction through the refractory walls can exceed 5 to 10% of the total heat release and can result in unsafe outside wall temperatures.

In refractory-lined systems where worker comfort or safety require cool outside wall temperatures, an analysis of wall heat loss is appropriate. Boundary conditions for the analysis would include:

Assume that the inside wall temperature will be maintained within a few hundred degrees of the hottest large areas or volumes in the combustion chamber. This assumption recognizes the fact that, since wall conduction is relatively poor, the inside temperature will approach radiative equilibrium with such intense radiation sources.

Allow for both convective and radiative losses from the outer wall. Rather than introducing the mathematical complexity of the fourth power of temperature dependency of radiation, use an overall natural convection plus radiation coefficient, available from standard texts [5]. Note, however, that if outside temperatures exceed, say, 200°C, a more careful consideration of radiation is appropriate.

For one-dimensional heat transfer by conduction in the x-direction, the heat transfer per unit area $\dot{Q}_a$ is given by:

$$\dot{Q}_a = -\lambda \frac{dT}{dx} \quad \text{kcal hr}^{-1} \text{m}^{-2} \tag{62}$$

where λ is the thermal conductivity and dT/dx is the thermal gradient. The minus sign acknowledges the fact that heat flow in the positive x-direction requires a decline in temperature with increasing x. The conversion factor for λ from the commonly tabulated values (e.g., Ref. 4) of Btu hr^{-1} ft^{-2} (°F/ft)$^{-1}$ to kcal hr^{-1} m^{-2} (°C/m)$^{-1}$ is 1.487.

For steady, one-dimensional conduction in isotropic solids, the thermal gradient will be linear and Eq. (62) becomes

$$\dot{Q}_a = -\lambda \frac{\Delta T}{\Delta x} \tag{63}$$

where Δx is the thickness of the slab and ΔT the surface to surface temperature difference.

It should be noted that the thermal conductivity often varies appreciably over the temperature extremes common to combustion systems, and a suitable average must often be selected. More complex problems which relax the assumptions indicated above (isotropic materials, one-dimensionality, etc.) may often be solved by the use of special texts (e.g., Ref. 64).

Example 13. A horizontal, cylindrical incinerator 4 m in diameter is burning liquid waste in a chamber with 20 cm thick insulating firebrick walls, sheathed with 6 mm thick steel. The mean radiative temperature of the flame is 1100°C. What is the approximate outside skin temperature?

Assume the equilibrium temperature of the inside wall is 1000°C and that the ambient is 25°C. Simplify the analysis by assuming that the thermal gradient within the wall and within the steel shell is linear. Figure 14 illustrates the situation.

$$\dot{Q}_a\Big|_R = \text{Conduction through refractory} = \frac{\lambda_R(T_1 - T_2)}{X_R} \text{ kcal hr}^{-1}\text{ m}^{-2}$$

$$\dot{Q}_a\Big|_S = \begin{array}{c}\text{Conduction through steel}\\ \text{(neglecting conduction}\\ \text{in air film)}\end{array} = \frac{\lambda_S(T_2 - T_3)}{X_S} \text{ kcal hr}^{-1}\text{ m}^{-2}$$

$$\dot{Q}_a\Big|_{a,c} = \text{Convection to ambient} = h_c(T_3 - T_a) \text{ kcal hr}^{-1}\text{ m}^{-1}$$

where

$$h_c = 2.84\left(\frac{\Delta T_s}{D_0}\right)^{0.25} \text{ kcal hr}^{-1}\text{ m}^{-2}\text{ °C} \tag{64}$$

Note that Eq. (64) is a dimensional equation, with ΔT_s being the temperature difference between surface and ambient in °C and D_0 is the outside diameter in meters. The equation applies to horizontal cylinders or to long vertical cylinders only. For other configurations:

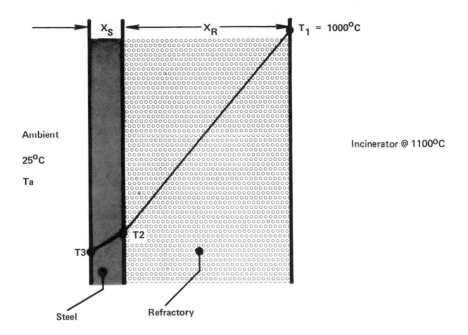

FIG. 14. Schematic of process in Example 13.

Vertical plates higher than 1 m:

$$h_c = 4.23 \, (\Delta T_s)^{0.25} \quad \text{kcal hr}^{-1} \, \text{m}^{-2} \, {}^{\circ}\text{C}^{-1} \tag{65}$$

Horizontal plates

    facing upward

$$h_c = 5.93 \, (\Delta T_s)^{0.25} \quad \text{kcal hr}^{-1} \, \text{m}^{-2} \, {}^{\circ}\text{C}^{-1} \tag{66}$$

    facing downward

$$h_c = 3.12 \, (\Delta T_s)^{0.25} \quad \text{kcal hr}^{-1} \, \text{m}^{-2} \, {}^{\circ}\text{C}^{-1} \tag{67}$$

$$\dot{Q}_a \Big|_{a,r} = \text{Radiation to ambient} = h_r (T_3 - T_a) \quad \text{kcal hr}^{-1} \, \text{m}^{-2} \tag{68}$$

where

$$h_r = \frac{4.92 \times 10^{-8}(T_{3'}^4 - T_{a'}^4)(\epsilon_s)}{(T_3 - T_a)} \quad \text{kcal hr}^{-1} \text{ m}^{-2} \text{ K}^{-1} \qquad (69)$$

$T_{3'}$ and $T_{a'}$ are the surface and ambient temperatures (respectively) expressed in degrees Kelvin and ϵ_s is the emissivity of the outer steel shell.

Property values of use are:

$$\lambda_R = 0.15 \quad \text{kcal hr}^{-1} \text{ m}^{-2} (^\circ\text{C/m})^{-1}$$
$$\lambda_S = 37 \quad \text{kcal hr}^{-1} \text{ m}^{-2} (^\circ\text{C/m})^{-1}$$
$$\epsilon_S = 0.8 \quad \text{(Oxidized steel, rough oxide coat)}$$

Note that

$$\dot{Q}_a\Big|_R = \dot{Q}_a\Big|_S = \dot{Q}_a\Big|_{a,c} + \dot{Q}_a\Big|_{a,r}$$

Rather than attemtping to solve a fourth order equation, assume T_3, calculate h_c and h_r, calculate T_3, and iterate to acceptable convergence:

Assume

$$T_3 = 120^\circ\text{C (393 K)}$$

$$h_c = 2.84\left(\frac{120 - 25}{4}\right)^{0.25} = 6.27 \text{ kcal hr}^{-1} \text{ m}^{-2} \, ^\circ\text{C}^{-1}$$

$$h_r = \frac{4.92 \times 10^{-8}(393^4 - 298^4)(0.8)}{(120 - 25)} = 6.61 \text{ kcal hr}^{-1} \text{ m}^{-2} \, ^\circ\text{C}^{-1}$$

$$h_{c+r} = h_c + h_r = 12.89 \text{ kcal hr}^{-1} \text{ m}^{-2} \, ^\circ\text{C}^{-1}$$

Eliminating T_2 between the steel-to-refractory and the refractory-to-ambient heat loss terms and using the notation

$$C_1 = 1 + \frac{\lambda_S X_R}{\lambda_R X_S} = 8223$$

$$C_2 = \frac{(X_R)}{\lambda_R} h_{c+r} = 17.19$$

Then

$$T_3 = \frac{(C_1 - 1)T_1 + C_1 C_2 T_a}{(C_1 - 1 + C_1 C_2)}$$

$T_3 = 78.6°C$ and $T_2 = 78.7°C$.

Clearly, the conductive resistance of the steel is negligible in comparison to that of the firebrick and the problem could be simplified by neglecting it.

Recomputing h_c and h_r for the new T_3 yields

$$h_c = 5.43 \text{ kcal hr}^{-1} \text{ m}^{-2} °C^{-1}$$

$$h_r = 5.43 \text{ kcal hr}^{-1} \text{ m}^{-2} °C^{-1}$$

$$h_{c+r} = 10.86 \text{ kcal hr}^{-1} \text{ m}^{-2} °C^{-1}$$

and

$$T_3 = 88°C$$

acceptable convergence.

It is noteworthy that changing the assumed inside wall temperature by 100°C would change T_3 by less than 10°C. Also, if the outside wall is as hot as 90°C (194°F), care should be exercised to prevent worker contact or consideration should be given to lagging the wall with insulation to bring the temperature within safe limits.

2. Convection

Convection, though important in boiler design, is usually less important than radiation in combustion system analysis. A particular exception occurs when jets of heated air or flames impinge upon a surface. This can happen when sidewall overfire air jets are discharged across narrow furnaces or burner flames impinge on the opposite end of the furnace. The heat transfer rates at the point of impingement can be extremely high. Consequent damages to refractory or to boiler tubes can be excessive. For this reason, flame length or jet penetration calculations should be used to assure a reasonable margin of safety.

3. Radiation

In high temperature combustors, radiation is the dominant mechanism for heat transfer. The relationship defining the total radiant emissive power

W_B from a black body at a temperature T is known as the Stefan–Boltzmann law:

$$W_B = \sigma T^4 \text{ kcal m}^{-2} \text{ hr}^{-1} \tag{70}$$

where

σ = the Stefan–Boltzmann Constant

$$= 4.88 \times 10^{-8} \text{ kcal m}^{-2} \text{ hr}^{-1} \text{ K}^{-4}$$

or

$$0.171 \times 10^{-8} \text{ Btu ft}^{-2} \text{ hr}^{-1} \text{ }^\circ\text{R}^{-4}$$

One should note the following:

The emissive power is spectral in nature (being different at different wavelengths). For black bodies (which absorb and emit fully at all wavelengths), this is unimportant. For solids, liquids, or gases which absorb or emit preferentially in one or more spectral regions, however, (importantly including carbon dioxide and water vapor) the spectral characteristics of both emitter and absorber must be taken into account. To simplify computation, the concept of a "grey body" has been developed. At the same temperature, the total emissive power of a grey body W is somewhat less than a black body. The ratio of emissive power of the grey to the black body at the same temperature is known as the emissivity, i.e.,

$$\frac{W}{W_B} = \epsilon \tag{71}$$

The emissivity of surfaces generally increases as tarnish or roughness increases (e.g., oxidized steel or refractory) and may change with temperature.

Radiation follows straight lines. Thus, the geometrical relationship between surfaces and volumes has a direct impact on the net flux. Also, surfaces will absorb and reradiate and may reflect heat, thus compounding the difficulty of analysis.

Gases and luminous flames are radiators. Although carbon dioxide and water vapor (particularly) are essentially transparent in the visible region of the spectrum, they absorb strongly in other spectral regions and participate significantly in radiant heat transfer in furnaces. It should be noted that the overall characteristics of gas emissivity (a "grey gas") may be conveniently approximated with an exponential formulation

$$\epsilon_{gas} = 1 - \exp(-apx) \tag{72}$$

where p is the partial pressure of the gas, x is the thickness of the gas layer, and a is a constant characteristic of the gas. Thus, thick sections of gas at high concentrations will have relatively higher emissivity than the converse. As an indication of the importance of these effects, at 1100°C, a 3-m thickness of water vapor at a partial pressure of 0.1 atm in air has an emissivity of 0.2, about 25% of that of a refractory wall.

Similarly, flames emit a significant radiative flux due both to the radiative emission of gases and to the hot particulate matter (ash and soot) contained within it.

For a thorough treatment of radiative transport, the serious student is directed to texts in the field [e.g., Ref. 27 and 63].

4. Heat Transfer Implications in Design

Although a detailed treatment of heat transfer in furnaces is beyond the scope of this chapter, several general observations should be made:

The designer should be aware of the heat requirement to initiate and maintain a steady burning condition. When heat recovery is of no interest, no serious compromises are required and the shape and operating temperature of the surfaces in radiative "touch" with the incoming waste can be kept such as to assure rapid and stable ignition. When heating of boiler surfaces or stock is an objective, care should be given to avoid robbing the feed zone of the heat flux required to maintain steady burning.

The designer should consider the heat fluxes within and through the furnace as they may affect wall temperatures. Often, in waste burning systems, the ash materials may react with the refractory at elevated temperatures to yield low melting eutectics. Such processes result in rapid, even catastrophic, wall degradation. Also, high wall temperatures may foster adherence of ash (slag build-up) which can cause mechanical damage to walls and, in mass burning systems, to the grates when the accumulated slag breaks off and falls. Such slag deposits can build up to the point where the flow of gases and even waste moving on grates can be impeded as, for example, by completely bridging across a 3-m wide furnace.

The designer should also be aware that cold surfaces are heat sinks, draining away the radiative energy from within a furnace. Many furnaces are constructed with (from a radiative heat loss standpoint) clear "views" of cold zones from the hot primary chamber. The result is a chilling of the combustion zone with noticeable increases in soot and carbon monoxide generation.

C. FLUID MECHANICS

Furnace fluid mechanics is complex, with flows driven by jets and by buoyancy and interacting in swirling recirculating eddys, all traversing complex geometrical sections. Yet a basic understanding of furnace flow processes gives great insight into the design and/or anticipation of air jets, burner-chamber interactions, mixing problems, and other effects which are vital parts of the combustion process.

This portion of the chapter is divided into two sections: driven flows (those directly at the command of the designer), and induced flows (those arising from driven flows and those induced by buoyancy effects).

1. Driven Flow

The designer has within his control an effective and flexible flow control tool: jets of air or steam and/or, for gaseous, liquid, or pulverized fuels, jets of fuel. These jetting flows serve to introduce fuel, introduce combustion or temperature control air or, particularly for the steam jet, serve to inject energy to mix and/or direct the furnace gas flow.

a. Jet Flow

The following analysis of jet flow is drawn heavily from the portions of Refs. 21 and 65 dealing with the interrelationship of furnace flow with pollutant emissions.

(1) Introduction

The gasification behavior of refuse beds and the need to induce mixing of furnace gases indicate that situations exist with incinerator furnaces where jet systems could be of assistance in realizing better burnout of combustible pollutants and in controlling furnace temperature distributions. A review of the design and operating characteristics of existing incineration systems [21] and discussions with incinerator designers suggest the need for better correlations supporting the design of these jet systems. As discussed below, the fluctuating conditions of gas movement and composition within incinerator furnaces present a considerable challenge to the detailed analysis of any device which interacts with the flow and combustion processes. As a consequence, the analysis is necessarily somewhat simplified. However, the results will support the design of practical systems which will perform effectively and in accord with the expectations of the designers [65].

(a) <u>Use of Jets</u>. Jets have been utilized for many years as an integral part of furnaces, boilers, and other combustion systems. In boilers fired with pulverized coal, for example, air jets are used to convey the fuel into the combustion chamber, to control the heat release patterns, and to supply secondary air for complete combustion. In processes employing a burning fuel bed, properly placed air jets supply secondary air where needed above the fuel bed to complete combustion. Also, jets of air and/or steam are used to induce turbulence and to control temperature by dilution of furnace gases.

The important characteristics of jets which underlie all of these uses are:

The controlled addition of <u>mass</u> to contribute to the oxidation process (air jets) or to serve as a thermal sink to maintain gas temperatures below levels where slagging, corrosion, or materials degradation may occur (air or steam jets).

The controlled addition of <u>momentum</u> to promote mixing of the jet-conveyed gas with gases in the combustion chamber or to promote mixing of gases from different parts of the combustion chamber. In the latter case, high-pressure steam jets are often used to provide high momentum fluxes with a minimum introduction of mass.

The basic challenge to the combustion system designer is to employ these characteristics to maximum advantage in meeting his overall design goals.

(b) <u>General Characteristics of Jets</u>. Because of the longstanding practical interest in the use of jets, a large body of literature has been developed which quantitatively characterizes the nature of jet flow. Jets are conveniently categorized, according to flow regime (laminar or turbulent, supersonic or subsonic) and geometry (round or plane). Laminar jets occur only at very low jet velocities and are of no interest here. Supersonic jets, for describing high pressure steam flows, are of potential interest but are not considered here. Plane jets, which issue from a slot finite in one dimension and effectively infinite in the other, are primarily of academic interest. We will, therefore, focus in this discussion on round, low subsonic, turbulent jets and return later to the fact that a row of closely spaced round jets behaves, to a degree, like a plane jet.

Other important parameters characterizing jet flow behavior include the relative densities of the jet and ambient fluids, the velocity of the ambient fluid relative to the jet velocity, and the degree to which the space into which the jet issues is confined by walls. Also, in situations where combustion can occur (jets of fuel into air as in burners and jets of air into fuel vapors, the so-called inverted flame), the initial temperature and combustible content of the jet and ambient fluid are of interest. All of these factors are important in the application of jets to incinerators, and their effects, singly and in combination, on jet characteristics are discussed. To set the stage for this discussion, we consider here the basic characteristics of jets issuing into an infinite atmosphere of quiescent fluid of the same density as the jet fluid.

The round, isothermal turbulent jet shows three characteristic regions (Fig. 15). Immediately adjacent to the nozzle mouth is the <u>mixing region.</u> Fluid leaves the nozzle with an essentially flat velocity profile. The large velocity gradients between this "potential core" and the ambient fluid induce turbulence which causes ambient fluid to mix into the jet. The mixing results in momentum transfer between the jet and ambient fluids and progressively destroys the flat velocity profile. At a distance of about 4.5 jet nozzle diameters downstream, the influence of shear forces reaches the centerline of the jet and eliminates the potential core [13].

It is important to note that the "nozzle diameter" characterizing jet flow is not necessarily the physical dimension of the orifice from which the jet issues. If, for example, the jet issues from a sheet metal plenum, a flow contraction to about 60% of the open discharge area (the area of the vena contracta) characteristic of the flow past a sharp edged orifice, will define the effective nozzle diameter and the location of the effective jet discharge plane will be displaced about two-thirds of a diameter downstream of the orifice (the location of the vena contracta). If a relatively long (2-3 diameters) constant area section lies upstream of the discharge plane, the nozzle diameter may be taken as the orifice diameter. Attention should be given, therefore, to the geometry of the entire nozzle fluid delivery system in analysis of jet behavior.

In the region from 4.5 to about eight diameters downstream, the transition of the flat entrance velocity profile to a fully developed profile is completed. Beyond this <u>transition region</u> the velocity profile retains a more or less constant shape relative to the velocity on the axis of the jet and is referred to as "self-preserving," the <u>fully developed</u> region.

Important jet characteristics include

1. The centerline velocity and concentration changes with axial distance from the nozzle mouth

2. The shape of the radial velocity and concentration profiles in the fully developed region

3. The intensity of turbulence in the jet

4. The rate of entrainment of ambient fluid into the jet

These characteristics are all interrelated; turbulence generated by high velocity gradients induces entrainment which causes momentum and mass

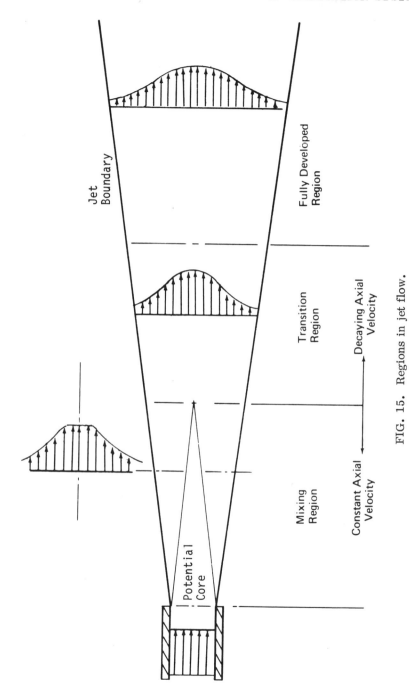

FIG. 15. Regions in jet flow.

transfer between the jet and the ambient fluid. These characteristics are important in practice because they determine the quantitative effect of firing a jet into a combustion chamber. The axial decay of velocity establishes how far the jet penetrates into the chamber. The radial velocity distributions determine how large a volume is affected by the jet. The entrainment rates determine how effectively furnace gases are mixed along the jet path.

(2) The Use of Jets for Combustion Control

Municipal incineration is an important source of combustible pollutants. Studies discussed in Ref. 21 showed that these pollutants would necessarily arise in the gasification zone of the grate and could possibly arise in the discharge zone.

These observations lead to the conclusion that systems are needed to provide air near the gasification zone and/or to induce high intensity turbulence at strategic locations within the incinerator furnace. Although passive mixing systems, such as baffles or checkerwork, may have some value in the inducement of turbulence, they clearly are not useful in supplying air and are not alterable to cope with changes in the distribution of combustible pollutant release along the bed or throughout the chamber.

(a) Jet Design for Incinerators—A Statement of the Problem. Combustible pollutants appear to be generated along the full length of a mass burning incinerator grate, although their discharge rate into the overfire volume is relatively low in the drying and ignition zones prior to the introduction of underfire air. From the standpoint of kilograms per hour per square meter release rate, the gasification zone probably qualifies as the single most important source of carbon monoxide, soot and hydrocarbons in the system. Carbon monoxide and coked ash material will be evolved in the region between the gasification zone and the burnout region. Overfire air is definitely required in the region of gasification and char burnout, reduced undergrate air flows and turbulence inducement is required in the area over the discharge grate, and some means may be required to increase the general level of turbulence throughout the upper regions of the incinerator furnace.

In incinerators for liquid or gaseous wastes, mixing processes often depend on jet systems, both to introduce the waste and to assure complete combustion. This latter concern is particularly important if the waste is a toxic material or includes pathogenic organisms.

The specification of jets for incinerator applications meeting the requirements listed above places great demands upon the designer. It is clear that the jet behavior should be known in a flow field where combustion, crossflow, and buoyancy effects are all potentially important, and, for some systems, the jets must operate over long distances. This latter characteristic arises from the shape of most continuous feed mass burning incinerators

which tend to be long and narrow. Thus, jets directed over the discharge
grate region which are expected to carry bed off-gases back towards the
pyrolysis region must act over distances of 3 to 10 m (20 to 100 or more jet
diameters). The location, number, and flow parameters appropriate to
these jets should be consistent with the overall furnace geometry, should be
easily maintained and operated, and should be controllable to the extent
demanded by the fluctuations in refuse composition and burning characteristics. Particularly, when jets are used for secondary air addition, the jet
design should add sufficient air to meet the oxygen requirement of the rising
fuel vapors, yet not provide so much air as to overly cool the gases, thus
quenching combustion. Also, the draft capabilities of the furnace must be
considered in determining the amounts of air introduced.

(b) Experience in Jet Application for Coal-Burning Systems. Overfire air
systems have been used for over 90 years in coal burning practice. In some
respects, the combustion characteristics of coal burning on a grate are
similar to those of refuse. Typically, however, coal ignites more readily
(partly due to its lower moisture content), burns with more regularity and
predictability, and, for overfeed or crossfeed situations, is typically burned
in furnaces with grates which are short relative to those used in many continuous feed refuse burning incinerators. Therefore, although the problems
are not identical, it is of value to review experience in coal burning practice
as an indication of the potential of jet systems for combustion control.

The use of controlled overfire air in industrial solid fuel combustion
systems was stimulated by the desire to improve boiler efficiency through
complete combustion of soot and carbon monoxide and to reduce smoke
emissions. Although the historical pattern of technological development of
overfire air systems is unclear, Stern [66] mentions that patents and active
marketing of steam-air jets, primarily for smoke control, began in 1880.
Quantitative appreciation of the benefits of smokeless combustion on overall fuel economy was widely argued until documented by Switzer [67] in 1910.
Switzer's work, carried out at the University of Tennessee, involved measurements of jet system steam consumption, smoke intensity and boiler efficiency on a hand fired return-tubular boiler fired with bituminous coal.
The results of his tests showed an increase in thermal efficiency from 52.6
to 62.1%, an increase in the effective range of the boiler from 80 to 105%
of its rated capacity before smoking occurred, and a steam consumption for
the overfire jet system of only 4.6% of the total steam raised.

Recognition of the importance of overfire air and mixing stimulated
considerable research in the first decade of this century. Some of the more
completely documented and detailed laboratory and field data was produced
by the Bureau of Mines who were conducting "investigations to determine
how fuels belonging to or for the use of the United States Government can
be utilized with greater efficiency." Kreisinger et al. [68] studied the

combustion behavior of several coals in a special research furnace under a
variety of combustion air and firing rate conditions. Their work showed a
strong relationship between the burnedness of the flue gas, the properties
of the coal, and the size of the combustion space (Fig. 16). Their results
were interpreted in agreement with prior suggestions of Breckenridge [69],
to result from differences in the emission of volatiles between coal varieties and the rate-limiting effects of inadequate mixing. Their correlating
parameter, which they called the "undeveloped heat of combustible gases,"
represented the remaining heat of combustion of the flue gases relative to
the heat of combustion of the coal burned. In view of the trends shown in
Fig. 16 and the composition data in Table 22 it would appear that the volume
requirements for complete burnout in refuse incineration may be considerably in excess of those acceptable in coal fired combustors. Quantitative
extrapolation from their data to incinerators, however, would be highly
speculative. Unfortunately, within the scope of the USBM experimental program, generalized design guides for the flow rate and locations appropriate
for overfire air systems were not developed.

Some of the earliest test work directly aimed at finding the benefits of
overfire air in utility combustion systems was conducted in 1926 by Grunert
[70] on forced draft chain grate stokers at the Commonwealth Edison Company in Chicago. Grunert's data showed that overfire jets discharging over
the ignition zone could reduce the carbon monoxide levels at the entrance to
the first pass of boiler tubes from an average value of 1% to essentially
zero. Also, the gas temperature and composition profile could be made
considerably more uniform. Of importance to fuel economy, it was found
that although additional air was introduced through the overfire air jets, the
total combustion air was susceptible to reduction. Similar work in Milwau-
kee reported by Drewry [71] also showed performance improvement (an increase of 7.2% in boiler efficiency), smoke elimination, and complete burnout of combustibles within the firebox. Once again, however, design correlations generally applicable to the coal-burning industry were not presented.

Major contributions to the overfire jet design art were published in the
mid-1930s. Of particular importance were reports on a number of meticu-
lous test programs carried out in Germany, perhaps typified by the work of
Mayer [72]. Although still not providing generalized design criteria, Mayer
made gas composition traverses (45 points) within a traveling grate stoker
furnace firing low-volatile bituminous coal. His results are shown in
Fig. 17.

As a measure of the completeness of combustion in the overfire space,
Mayer determined the heating value of the gases ($kcal/m^3$) as calculated
from the complete gas analysis. Without overfire air, as seen in Fig. 17a,
strata of combustible gases rise into the combustion chamber and persist
as the gases enter the first boiler pass. This is indicated by the zero
heating value curve which is not closed. Figure 17b shows the effect of

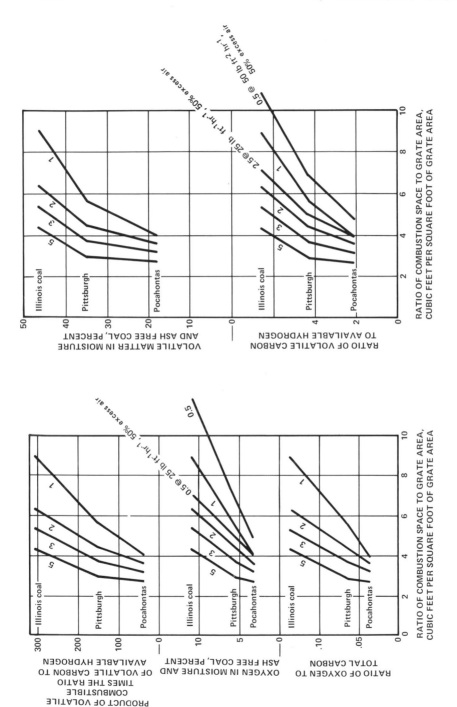

FIG. 16.   Relation between coal characteristics and size of combustion space required in USBM test furnace at combustion rate of 50 lb hr^{-1} ft^{-2} and 50% excess air. Numbers on graphs are the percentage of the heat of combustion of the original coal which appears as unreleased heat of combustion in the furnace gases. (From Ref. 68.)

TABLE 22

Chemical Characteristics of Coal and Refuse[a]

Item	Characteristics[b]	Pocahontas coal	Pittsburgh coal	Illinois coal	Refuse
1	Volatile matter	18.05	34.77	46.52	88.02
2	Fixed carbon	81.95	65.23	53.48	11.99
3	Total carbon	90.50	85.7	79.7	50.22
4	Volatile carbon (item 3 to item 4)	8.55	20.47	26.22	38.22
5	Available hydrogen	3.96	4.70	3.96	1.57
6	Ratio volume C to available H_2	2.16	4.35	6.60	24.34
7	Oxygen	3.32	5.59	10.93	41.60
8	Nitrogen	1.19	1.73	1.70	1.27
9	Percentage of moisture accompanying 100% of M & AF coal or refuse	2.53	2.88	22.07	55.19
10	Product of items 1 and 6	39	151	307	2142
11	Ratio of oxygen to total carbon	0.0367	0.0652	0.137	0.828
12	Total moisture in furnace per kilogram of coal or refuse reduced to M & AF basis (kilogram)	0.409	0.501	0.700	1.161

[a] Data on coal are from Ref. 68; data on refuse are from Ref. 21.
[b] Items 1 through 8 and 11: percent on moisture and ash-free basis (M & AF).

medium-pressure overfire air jets. It can be seen that combustion is improved, yet some fraction of the combustible still enters the boiler passes. Figure 17c shows the effect of further increases in the overfire air plenum pressure. Under these latter conditions, combustion is complete within the furnace volume. For these tests, Mayer employed jets directed towards the bed just beyond the ignition arch.

Also in the 1930s, developments in fluid mechanics by Prandtl and others provided mathematical and experimental correlation on the behavior of jets. Application of this understanding to furnace situations was presented in some detail by Davis [73]. His correlations, although based on

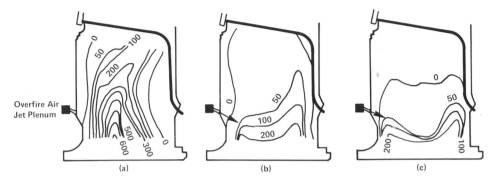

FIG. 17. Lines of equal heating value (kilocalories per standard cubic meter) of flue gas firing low-volatile bituminous coal at a fuel rate of 137 kg m^2 hr^{-1}: (a) No overfire air, (b) overfire air pressure 13 mm Hg (7 in. H$_2$O), (c) overfire air pressure 18.7 mm Hg (10 in. H$_2$O). (From Ref. 72.)

greatly simplified assumptions, were of considerable interest to furnace designers at that time. As an example of the applicability of his work, Davis explored the trajectories anticipated for jets discharging over coal fires and compared his calculated trajectories with data by Robey and Harlow [74] on flame shape in a furnace at various levels of overfire air. The results of this comparison are shown in Fig. 18. Although general agreement is shown between the jet trajectory and the flame patterns, correlation of the meaning of these parameters with completeness of combustion is unclear and not supported by Robey and Harlow's data. The results do give confidence, however, that jet trajectories can be calculated under a variety of furnace conditions to produce reasonable estimates of behavior and thus permit avoidance of impingement of the jet in the bed.

Although work on the applications and advantages of overfire air continued through World War II, particularly with reference to avoidance of smoke in naval vessels to preclude easy identification and submarine attack [66], the wide introduction of pulverized coal firing in electric utility boilers and the rapid encroachment of oil and gas into the domestic, commercial, and industrial fuel markets rapidly decreased the incentive for continued research into overfire jet systems for application in stoker fired combustors. Indeed, the number of literature references on this topic fall off rapidly after 1945.

In summary, during the 70 or more years during which overfire air jet application to stoker fired systems was of significant importance, no generalized design criteria of broad applicability had been developed. An art had arisen regarding the use of overfire jets, typically over the ignition arch and in the sidewalls of traveling grate stokers burning bituminous coal. Sufficient jet design technology had been developed to allow specification of jets which would penetrate adequately the upflow of gases arising from the bed and which served to smooth the temperature and gas composition profiles at the entrance to the boiler passes above. Even in 1951, however, the comment was made by Gumz [75], a well-recognized contributor to combustion technology, that "the number of nozzles, their location and direction are the most disputed factors in the use of overfire air jets."

(3) Jet Dynamics and Design Guidelines

In 1880, the first patents were issued for steam-air jet devices to supply overfire air and induce turbulence in hand fired furnaces burning bitumi-

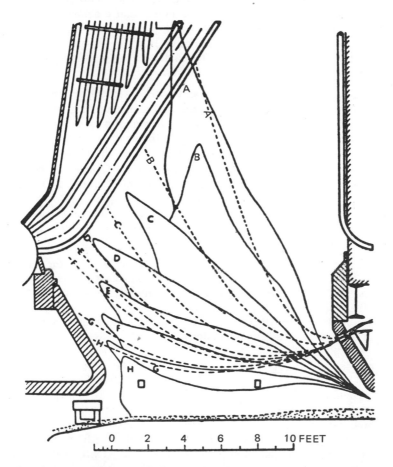

FIG. 18. Comparison of observed flame contours and calculated trajectories of overfire air jets. Percentage of overfire air at the following points: A, 565; B, 10.5; C, 16.6; D, 20.0; E, 21.4; F, 22.8; G, 26.8; H, 28.8. (From Ref. 73.)

nous coal. The development and marketing of these proprietary jet systems reflected a need for improvements in combustion efficiency and for means to reduce smoke emissions. Since that time, a number of refinements in the physical arrangements and design characteristics of overfire jets have been offered to the technical community. However, few instances can be found where comprehensive design correlations are presented in the literature. In the great majority of cases (e.g., Refs. 76-78), the technical content of the papers is limited to documentation of improvements in performance (particularly with reference to smoke abatement) resulting from the use of specific arrays of overfire jets in a specific combustor. Most published design information deals with such topics as the pumping efficiency or the estimation of steam consumption in steam ejectors. One can find few instances where attempts were made to couple an analysis of jet behavior to an analysis of furnace behavior.

To some extent, the tendency of early workers to report only empirical results reflected the limitations of theoretical understanding or mathematical treatment techniques of their time. Also, the complexity of furnace dynamics presents a considerable challenge to the analyst and thus generalization is difficult. It is noteworthy, for example, that the rigorous mathematical treatment of the behavior of two-dimensional plane jets has only recently been solved in detail [79]. Solution of this problem required the use of high-speed computers and complex numerical techniques. Efforts at a similar analysis of the axisymmetrical round jet are now in process but, at the present time, trial solutions exceed reasonable core storage and computation time on the fastest, modern computers.

Ideally, the design basis used in overfire air jet designs for incinerator applications should recognize the effects of buoyancy, crossflow, and combustion as they are experienced in real incineration systems. The work of Davis [80] and Ivanov [81], though concerned with coal fired systems, produced correlations resting heavily on experimental results which provide a starting point for a design methodology.

(a) <u>Round Isothermal Jets</u>. The behavior of circular jets discharging into a quiescent, nonreacting environment at a temperature similar to that of the jet fluid provides the starting point in any review of jet dynamics. Indeed, the behavior of jets under such conditions has often been the primary guideline in the design of overfire jets for incinerator applications [82]. Because of the relatively simple nature of jet structure under such conditions, this configuration is perhaps the most studied, both analytically and experimentally, and good correlations are available describing jet trajectory, velocity, entrainment, turbulence levels, and the like.

The correlation of data taken by many experimenters leads to the following expressions for the axial decay of centerline velocity and concentration [13]

$$\frac{\bar{u}_m}{\bar{u}_0} = 6.3 \left(\frac{\rho_0}{\rho_a}\right)^{1/2} \frac{d_0}{(x + 0.6d_0)}$$ (73)

$$\frac{\bar{c}_m}{\bar{c}_0} = 5.0 \left(\frac{\rho_0}{\rho_a}\right)^{1/2} \frac{d_0}{(x + 0.8d_0)}$$ (74)

where $\bar{u}_m$ and $\bar{c}_m$ are the time averaged centerline velocity and concentration of jet fluid at distance x downstream from the nozzle mouth; $\bar{u}_0$ and $\bar{c}_0$ are the comparable characteristics at the nozzle mouth; and d_0 is the nozzle diameter. ρ_0 and ρ_a are the densities of the jet and ambient fluids, respectively. These equations apply only in the fully developed region (i.e., $x/d_0 > 8$) and are confirmed by other investigators (e.g., Ref. 83).

The experimentally measured radial velocity and concentration profiles in the fully developed region can be represented by either Gaussian or cosine functions. The Gaussian representations [13] are

$$\frac{\bar{u}}{\bar{u}_m} = \exp\left[-96\left(\frac{r}{x}\right)^2\right]$$ (75)

$$\frac{\bar{c}}{\bar{c}_m} = \exp\left[-57.5\left(\frac{r}{x}\right)^2\right]$$ (76)

where $\bar{u}$ and $\bar{c}$ are the time averaged velocity and concentration at distance x downstream and distance r from the jet centerline.

The spread of the jet is defined in terms of the half-angle to the half-velocity point (i.e., the angle subtended by the jet centerline and the line from the centerline at the nozzle mouth to the point where the velocity is one-half of the centerline velocity. This angle is independent of distance from the nozzle mouth in the fully developed region, a consequence of the self-preserving nature of the velocity profile. The half-angle of the half-velocity point is approximately $4.85°$, based on concentration in the same way, the half-angle is $6.2°$ [13].

The turbulent intensity of the jet is defined in terms of u' and v', the r.m.s. fluctuating velocity components in axial and radial directions, respectively. Data of Corrsin [84] show that the intensity ratio $u'/\bar{u}_m$ and $v'/\bar{u}_m$ depend on the ratio r/x. At $x/d_0 = 20$, each velocity ratio varies from about 27% at the centerline to about 5 to 7% at $r/x = 0.16$.

Ricou and Spaulding [85] measured entrainment rates and determined that the mass flow rate ($\dot{m}_x$) in the jet is linearly related to x according to

$$\frac{m_x}{m_o} = 0.32 \left(\frac{\rho_a^{'}}{\rho_0}\right)^{1/2} \left(\frac{x}{d_0}\right) \tag{77}$$

This relationship holds for all values of nozzle Reynolds number greater than 2.5×10^4 and for $x/d_0 > 6$. At $x/d_0 < 6$, the entrainment rate per unit jet length is lower, increasing progressively with distance until it stabilizes at the constant value corresponding to Eq. (77). Entrainment is an inherent characteristic of jet flows. Equation (77) indicates that the magnitude of the entrained flow exceeds the nozzle flow in only a few nozzle diameters. The importance of the entrainment phenomena is felt most strongly in ducted flows (e.g., an axially mounted burner discharging into a cylindrical chamber) when an insufficiency of fluid relative to the entrainment requirement leads to significant recirculation flows (see Section C.2)

The relationships describing the behavior of isothermal jets entering a quiescent fluid are well documented and form the basis for design criteria relative to the use of jets in incinerators. In this application, however, the effects of crossflow of the ambient fluid and density differences between the jet and ambient fluids are important. These effects are less well documented, and their inclusion in the design criteria poses some difficult problems. These matters are dealt with below.

(b) <u>Bouyancy Effects</u>. When the jet and ambient fluids are of different density, the buoyant forces acting on the jet can cause deflections of the jet trajectory. This effect is potentially important in incinerator applications since the air introduced by the jets will be much colder than the furnace gases and hence of higher density. From an incinerator design and operating standpoint, this could be critical: jets could "sink" from an anticipated flow trajectory passing above the bed to one causing entrainment of particulate from the bed or causing overheating of the grates with a "blowpipe" effect.

Relatively little experimental or theoretical work has been done to characterize jet performance under these conditions. Figure 19 shows the geometry of the system considered and defines symbols used in the discussion. A jet of density ρ_0 issues at a velocity of u_0 from a circular nozzle of diameter d_0. The ambient fluid is at rest and of density ρ_a.

Abramovich [83] analyzed the trajectory of a heated jet issuing into a cold ambient fluid and compared his theoretical result with the data of Syrkin and Lyakhovskiy [86]. The resulting expression is

$$\left(\frac{y}{d_0}\right) = 0.052 \left[\frac{g\, d_0}{\overline{u}_0^2}\right] \frac{\rho_a - \rho_0}{\rho_0} \left(\frac{x}{d_0}\right)^3 \tag{78}$$

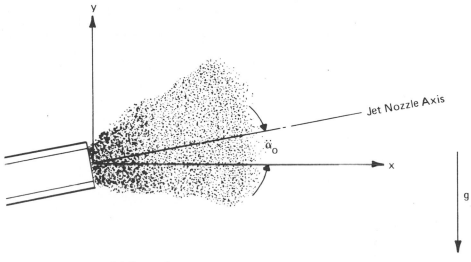

Jet Properties:
    Nozzle Velocity u_0
    Density ρ_0

Ambient Fluid Properties
    At Rest
    Density ρ_a

FIG. 19. Schematic of jet flow.

Figure 20 shows a comparison of this expression with experimental data in which the ratio $(\rho_a - \rho_0)/\rho_0$ was varied in the approximate range of 0.2 to 0.8. Equation 78 generally underestimates the buoyancy-induced deflection of the jet.

Field et al. [13] considered the behavior of a buoyant jet and obtained the expression

$$\frac{y}{d_0} = \left(\frac{x}{d_0}\right) \tan \alpha_0 + \frac{0.047}{\cos \alpha_0} \left(\frac{g \, d_0}{\bar{u}_0^2}\right) \left(\frac{\rho_a - \rho_0}{\rho_0}\right) \left(\frac{\rho_a}{\rho_0}\right)^{1/2} \left(\frac{x}{d_0}\right)^3 \tag{79}$$

For a jet injected normal to the gravity field ($\alpha_0 = 0$), Eq. (79) reduces to

$$\frac{y}{d_0} = 0.047 \left(\frac{g \, d_0}{\bar{u}_0^2}\right) \left(\frac{\rho_a - \rho_0}{\rho_0}\right) \left(\frac{\rho_a}{\rho_0}\right)^{1/2} \left(\frac{x}{d_0}\right)^3 \tag{79a}$$

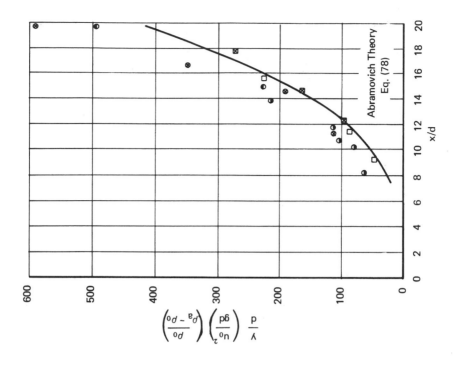

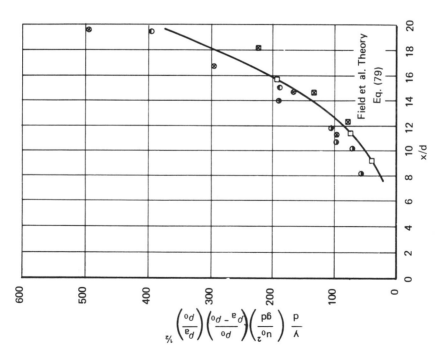

FIG. 20. Comparison of the predictions of Abramovich and Field et al. with the data of Syrkin and Lyakhousky on buoyant jet behavior. (From Refs. 13, 83, and 86.)

which differs from Eq. (78) in the value of the leading constant (0.047 as opposed to 0.052) and in the presence of the term $(\rho_a/\rho_0)^{1/2}$. In incinerator applications, where the jet and ambient temperatures are approximately 40°C (313 K) and 1430°C (1703 K), respectively, this term has the value of

$$(\rho_a/\rho_0) = (T_0/T_a)^{1/2}$$

$$= (313/1703)^{1/2}$$

$$= 0.43$$

The deflections predicted by the two equations will differ by a factor of two.

Figure 19 shows a comparison of Eq. (79a) with the data of Syrkin and Lyakhovskiy [86]. In these data, the term $(\rho_a/\rho_0)^{1/2}$ varies from about 1.1 to 1.4. The inclusion of term $(\rho_a/\rho_0)^{1/2}$ results in better agreement with the data, particularly at the larger values of x/d_0.

(c) Crossflow Effects. The need to understand the behavior of a jet issuing into a crossflow normal to the jet axis arises in the analysis of furnaces, plume dispersion from chimneys, and elsewhere. The deflection of the jet by the crossflow has been studied extensively, both experimentally and analytically, although most workers have limited their work to descriptions of centerline trajectory and gross entrainment rates. A recent review of the literature [55] indicates that no analysis has been carried out, either experimentally or theoretically, which characterizes in detail the radial distribution of velocity or concentration.

With crossflow, the interaction of the flow deflects the jet and alters the cross-sectional shape of the jet. Figure 21 is a diagram of the jet cross-section several nozzle diameters along the flow path. The originally circular cross-section has been distorted into a horseshoe shape by the shearing action of the external flow around the jet, and internal patterns of circulation have been set up. Measurements in the external flow around the jet show a decreased pressure downstream of the jet, recirculation of the external fluid, and a process leading to the periodic shedding of vortices into the wake of the jet. These phenomena are similar to those observed in the wake of a solid cylinder exposed to crossflow.

Dimensional analysis considerations suggest that the coordinates of the jet axis $(x/d_0, y/d_0)$ should depend on the ratio of momentum fluxes in the external and jet flows

$$M = \left(\frac{\rho_a u_1^2}{\rho_0 u_0^2}\right)$$

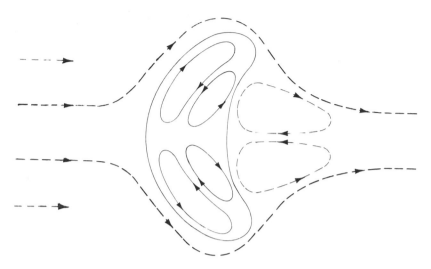

FIG. 21. Jet cross-section and circulation patterns for round jets in crossflow.

and the Reynolds number

$$Re = \left(\frac{\rho_0 \bar{u}_0 \, d_0}{\mu_0} \right)$$

For turbulent jets in the Reynolds number range above 10^4, correlation of experimental data suggests that the Reynolds number effect is negligible and the momentum ratio is the predominant variable characterizing the flow.

In terms of the geometry illustrated in Figure 22 the following expressions for computing the axial trajectory of a single jet have been reported. The jet axis is taken to be the locus of maximum velocity [13, 81, 83, 87–89].

$$\frac{y}{d_0} = 1.0(M)^{1.12} \left(\frac{x}{d_0} \right)^{2.64} \qquad 0 < M \le 0.023, \quad \alpha_0 = 0 \tag{80}$$

$$\frac{y}{d_0} = M \left(\frac{x}{d_0} \right)^{2.55} + (1 + M)(\tan \alpha_0) \left(\frac{x}{d_0} \right) \qquad 0.046 \le M \le 0.5 \tag{81}$$

$$\frac{y}{d_0} = (M)^{1.3} \left(\frac{x}{d_0} \right)^{3} + (\tan \alpha_0) \left(\frac{x}{d_0} \right) \qquad 0.001 \le M \le 0.8 \tag{82}$$

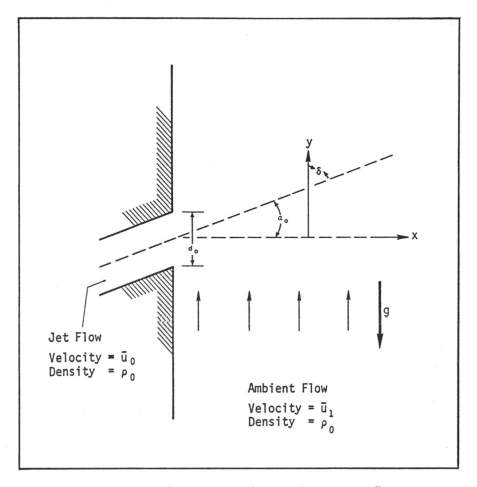

FIG. 22. Coordinate system for round jet in crossflow.

$$\frac{y - y^+}{d_0} = 5.5M^{1.175}\left(\frac{x - x^+}{d_0}\right)^{2.175} \qquad 0.01 \leq M \leq 0.028, \ \alpha_0 = 0 \qquad (83)$$

where y^+/d_0 and x^+/d_0 denote the end of a zone of establishment. Values vary somewhat with M but are of the order of one or less.

Abramovich [83] derived an analytical relation of the form:

$$\frac{x}{d_0} = \sqrt{\frac{39}{MC_x}} \ \ln\left[\frac{10 + (y/d_0) + [(y/d_0)^2 + 20(y/d_0) + (7/MC_x)\tan^2\alpha_0]^{1/2}}{10 + \sqrt{(7/MC_x)}\tan\alpha_0}\right]$$

$$(84)$$

where C_x is an effective drag coefficient relating drag on the jet to the momentum flux in the external flow and is approximately 3.

A simplified treatment of jet behavior in crossflow with temperature effects was presented by Davis [80] in 1937. The crossflow effect was introduced by the assumption that the jet rapidly acquired a velocity component equal to the crossflow velocity. The jet was then seen to follow a path corresponding to vector addition of the crossflow velocity to the jet centerline velocity (the latter being calculated using a simplified velocity decay law by Tollmien [90]. Temperature effects were introduced as being reflected in increases in jet velocity due to expansion of the cold nozzle fluid (initially at T_0) after mixing with the hot furnace gases (at T_a). Davis' final equation for the deflection (y) is given by:

$$y = \frac{u_1 x(x + 4d_0 \cos\alpha_0)}{2a_1 d_0 \bar{u}_0 \cos^2\alpha_0}\left(\frac{T_0}{T_a}\right)^{1/3} + x\tan\alpha_0 \qquad (85)$$

where a_1 is a constant depending on nozzle geometry (1.68 for round jets and 3.15 for long narrow plane jets). The many rough assumptions in Davis' analysis (some of which have been shown to be in error) would indicate that its use should be discouraged. Comparison of calculated trajectories with observed flame contours (Fig. 18) suggests it may have some general value. Interpretation of the meaning of the general agreement between calculated jet trajectory and flame contour as shown in Fig. 18 is uncertain, however, and use of the Davis equation in incinerator applications is questionable.

Patric [87] reported the trajectory of the jet axis (defined by the maximum concentration) to be

$$\frac{y}{d_0} = 1.0M^{1.25}\left(\frac{x}{d_0}\right)^{2.94} \qquad (\alpha_0 = 0) \tag{86}$$

Figure 23 shows a plot of the velocity axis [Eq. (81)] and the concentration axis [Eq. (86)] for a value of M = 0.05. The concentration axis shows a larger deflection than does the velocity axis. This is probably due, in part, to the assymetry of the external flow around the partially deflected jet. Also, recent calculations by Tatom [79] for plane jets suggest that under crossflow conditions, streamlines of ambient fluid can be expected to cross the jet velocity axis.

For our purposes, we are interested in jets which penetrate reasonably far into the crossflow (i.e., those which have a relatively high velocity relative to the crossflow). The empirical equations of Patrick [87] [Eq. (80)] and Ivanov [81] [Eq. (82)] were developed from data which satisfy this condition. Figures 24 and 25 show comparisons of these two equations for values of M of 0.001 ($\bar{u}_0/u_1 = 30$) and M = 0.01 ($\bar{u}_0/u_1 = 10$). Ivanov's expression predicts higher deflections at large (x/d_0, particularly at M = 0.01).

Ivanov [81] also investigated the effect of the spacing between jets (s) in a linear array on jet trajectory. He measured the trajectories of jets under conditions where M = 0.01 for s = 16, 8, and 4 jet diameters. His results are shown in Fig. 26 along with the trajectory of a single jet (infinite spacing). The data show that reducing the spacing between jets causes greater deflection of the jets. As spacing is reduced, the jets tend to merge into a curtain. The blocking effect of the curtain impedes the flow of external fluid around the jets and increases the effective deflecting force of the external fluid. The increase in deflection is most notable as s/d_0 is reduced from 16 to 8. Above $s/d_0 = 16$, the merging of the jets apparently occurs sufficiently far from the nozzle mouth to have little effect on the external flow. At $s/d_0 = 8$, the jet merge apparently takes place sufficiently close to the nozzle mouth that further reduction in spacing has little added effect.

Earlier data (Abramovich [83]) on water jets colored with dye issuing into a confined, crossflowing stream were correlated in terms of jet penetration distance. The penetration distance L_j was defined as the distance between the axis of the jet moving parallel to the flow and the plane containing the nozzle mouth. The axis was defined as being equidistant from the visible boundaries of the dyed jet. The resulting correlations was

$$\frac{L_j}{d_0} = k\,\frac{\bar{u}_0}{u_1} \tag{87}$$

where k is a coefficient depending on the angle of attack and the shape of the nozzle. Defining the angle of attack δ as the angle between the jet and

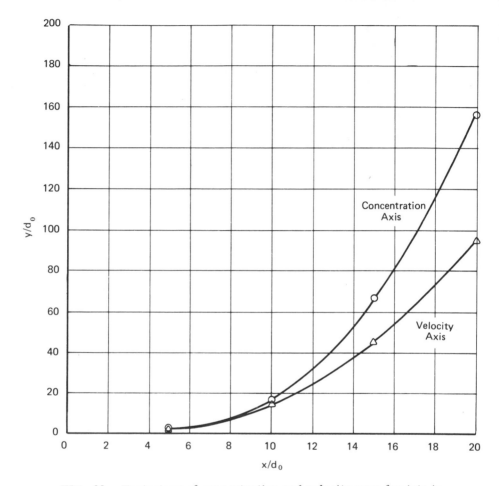

FIG. 23. Trajectory of concentration and velocity axes for jets in crossflow, M = 0.05 (data of Patrick, Ref. 87).

the crossflow velocity vectors and equal to $(90 - \alpha_0)$ in the terminology shown in Fig. 22, the recommended values of k are

For $\delta = 90°$ for round and square nozzles $k = 1.5$

For $\delta = 90°$ for rectangular nozzles $k = 1.8$

For $\delta = 120°$ for all nozzles $k = 1.85$

Figure 27 shows a comparison of Ivanov's correlation [Eq. (82)] with the jet penetration correlation [Eq. (87)], for M = 0.001 and M = 0.01.

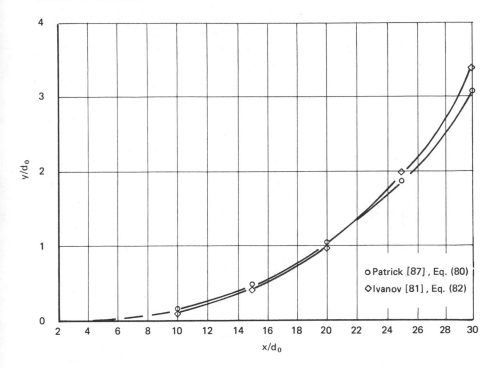

FIG. 24.   Comparison of trajectories at M = 0.001 (u_0/u_1 = 31.6) for jets in crossflow.

Equation (87) predicts a smaller jet penetration than does Eq. (82). There are two possible explanations for this discrepancy.

1. The data on which Eq. (82) was based do not extend to large values of x/d_0, and extrapolation of the data may be in error.

2. The data on which Eq. (87) was based were taken in a confined crossflow in which the lateral dimension (normal to both the jet axis and the crossflow) was sufficiently small to interfere with normal jet spreading. The jet effectively filled the cross-section in the lateral dimension, behaving like a series of jets at low spacing.

Ivanov's data (Fig. 26) show that penetration is reduced at lower spacing. The penetration given in Eq. (82) for M = 0.01 was reduced by 25% (see dotted trajectory marked s/d_0 = 4, M = 0.01 in Fig. 27). Agreement between this adjusted trajectory and the penetration given by Eq. (87) is better.

(d) Buoyancy and Crossflow. When a cold air jet is introduced into a crossflowing combustion chamber, both buoyancy and crossflow forces act simultaneously on the jet. Abramovich [83] reports the results of experiments

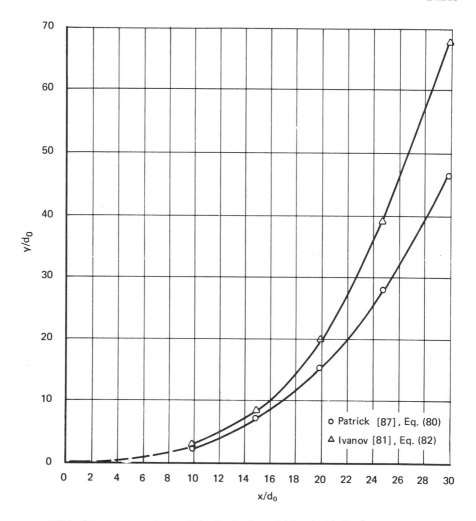

FIG. 25. Comparison of trajectories at $M = 0.01$ $(\overline{u}_0/u_1 = 10)$ for jets jets in crossflow.

conducted by injecting cold jets into a hot crossflow. Temperature ratios of as much as 3 to 1 were used (corresponding to the jet fluid having a density three times that of the crossflowing fluid), with the values of M in the range of 0.045 to 0.5. The normal crossflow trajectory equation correlated the data when the value of M was computed using actual fluid densities. From these data, Abramovich concluded that buoyancy effects could be neglected, other than as density differences were incorporated into the crossflow parameter M.

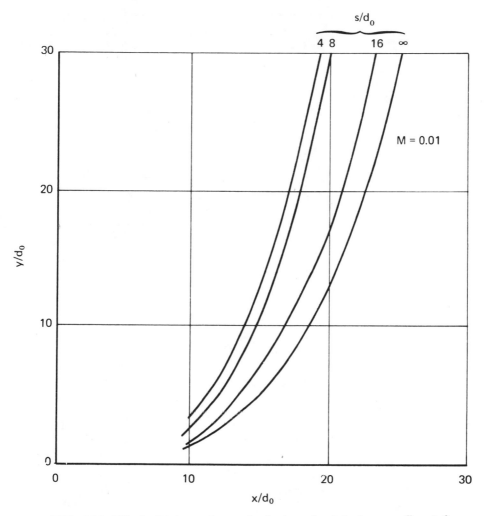

FIG. 26. Effect of jet spacing on trajectory for jets in crossflow (after Ivanov, Ref. 81).

The same conclusion was drawn by Ivanov [81], who injected hot jets into a cold crossflow. The ratio of temperature (and density) between jet and ambient fluids was 1.9, and M ranged from 0.005 to 0.02. The geometry of the tests was not clearly stated by either Abramovich or Ivanov. It appears that the buoyancy force acted in the same direction as the crossflow force in Ivanov's tests (i.e., a hot jet discharging into an upflow).

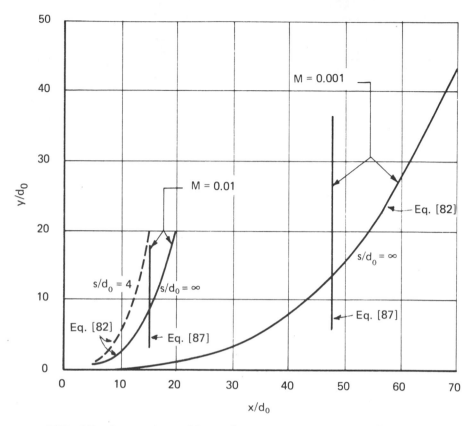

FIG. 27. Comparison of Ivanov's trajectory correlation [Eq. (82)] with jet penetration correlation [Eq. (87)]. (From Ref. 81.)

Application of these conclusions to incinerator design practice, how-ever, is subject to question because of the large geometrical scale-up in-volved. Physical reasoning suggests that the ratio of buoyant force to drag force acting on a nonisothermal jet in crossflow depends on scale. The buoyant force F_b is a body force and is, therefore, proportional to jet vol-ume. The drag force F_d exerted by the crossflow has the characteristics of a surface force and is, therefore, proportional to the effective cylindrical area of the jet. Therefore, the ratio of buoyant force to drag force is pro-portional to jet diameter. A simple analysis, discussed in Ref. 63 gives

$$\frac{F_b}{F_d} = \frac{\pi}{2C_x} \left(\frac{gd_0}{u_1^2}\right)\left(1 - \frac{\rho_0}{\rho_a}\right) \tag{88}$$

where C_x is the effective drag coefficient.

The value of the effective drag coefficient C_x is believed to be in the range of 1 to 5; analysis given in Ref. 63 suggests that 4.75 is an acceptable value.

Typical values of the physical parameters in Ivanov's experiments [81] are

$$d_0 = 5 \text{ to } 20 \text{ mm}$$

$$\frac{\bar{u}_0}{u_1} = 10 \text{ to } 20$$

$$\frac{T_0}{T_a} = \frac{\rho_a}{\rho_0} = 2$$

$$\frac{\rho_0 \bar{u}_0^2}{\rho_a u_1^2} = 100 \text{ to } 200$$

$$u_1 = 3.68 \text{ to } 4.16 \text{ m/sec}$$

The maximum value of F_b/F_d results from the maximum value of nozzle diameter d_0 and the minimum value of crossflow velocity u_1. Substitution of these values into Eq. (88) yields a force ratio of 0.003 which indicates that the crossflow effect completely dominated the buoyant effect on the small jets used in Ivanov's tests. Therefore, one would speculate that his results (no buoyancy effect) could be anticipated under his test conditions.

In incinerator applications, jet diameters in the neighborhood of 0.1 m are commonplace, along with crossflow velocities in the order of 0.6 to 1.5 m/sec. For a 0.1 m jet in a 0.6 m/sec crossflow with the same 1.91 density ratio as in Ivanov's tests,

$$\frac{F_b}{F_d} = \frac{(3.14)}{(2)(4.75)} \frac{(9.81)(0.1)}{(0.6)^2} \left(1 - \frac{1}{1.91} \right)$$

$$= 0.43$$

which suggests that the buoyant and crossflow forces are of the same order of magnitude.

(e) Design Methods. The correlations given above provide the basic tools for the analysis of jet behavior in real furnace environments. In general, the correlations are based in theory and corroborated with data. Translation of behavioral relationships into designs can be approached in two ways:

1. Detailed analysis of the actions and interactions of each component
    of the system under design, "building" an understanding of system
    behavior from an understanding of its parts

2. Assembly of generalized correlations into "rules of thumb" and the
    like which show applicability to a number of systems similar to the
    device in question

The jet design correlations above, when coupled with the bed burning and
chamber flow analyses, are supportive of the first approach. As such,
they are broadly applicable but their use makes demands upon the designer
for data and understanding which he may not possess.

Approaches to a rule of thumb, generalized method for overfire air
jet design have been proposed by Ivanov [81] and by Bituminous Coal Research, Inc. [91]. Although these design guides were developed for coal
fired boilers, they are presented here as an indication of approaches successful in other applications. Their applicability to incineration systems,
however, has been demonstrated only in one installation [30].

(i) The Method of Ivanov. Ivanov conducted a number of experiments in a
noncombusting model furnace to determine the effects of various jet configurations on the temperature profiles above a burning coal bed. The system being studied involved a rectangular furnace through which a single,
flat chain grate stoker drew a level fuel bed into the furnace at the front wall
and discharged ash just outside the rear wall. He concluded that:

It is preferable to position overfire jets in the front wall of the
furnace rather than in the rear wall.

Close spacing of the jets is desirable in order that the jets form
an effective curtain. Above this curtain, a rotary motion of the gases
is induced, which contributes greatly to the mixing process.

If maximum temperatures occur near the center of the grate,
rather than near the front, the design depth of penetration of the jet
should be increased by 5%.

Slightly better mixing is obtained if jets are fired from one wall,
rather than if the same flow is divided between jets on opposite walls,
whether the opposing sets of jets are directly opposed, staggered but
on the same horizontal level, or on different horizontal levels.

A given level of mixing is achieved at lower power cost and with
less air if small diameter jets are used rather than large ones.

For conditions which gave good mixing, Ivanov computed the jet penetrations
from Eq. (89).

$$L_j = d_0 k \left(\frac{\bar{u}_0}{u_1} \right)$$ (89)

and normalized the values obtained with respect to L_T, the axial length of the model furnace.

Using the values of k given on Page 150 to compute L_j, he correlated L_j/L_T with the relative jet spacing s/d_0 (where s is the center-to-center jet spacing and d_0 is the jet diameter) and obtained the following values:

(s/d_0)	Front arch furnace (L_j/L_T)	Rectangular furnace (L_j/L_T)
4	0.90	0.80
5	0.95	0.90
6	1.10	1.0

This correlation is the basis for his design method. Ivanov's design method is as follows:

Nozzles should be located not less than 1 nor more than 2 meters above the fuel bed.

The angle of inclination of the jets is determined by aiming the jets at a point on the grate 1.2 to 2 m from its far end. Jets fired from the underside of a front arch may be angled downward as much as 50° from the horizontal if the fuel bed is not disturbed by the resulting jet.

The relative jet spacing should be in the range of $s/d_0 = 4$ to 5.

The velocity of gases in the furnace at the cross-section where the jets are located is computed from known overall air rates and grate areas and corrected for temperature. This velocity is the crossflow velocity u_1 and the density is ρ_a.

The jet velocity is set by the capability of the overfire fan, but should always be 60 m/sec for cold jets and 70 m/sec for heated jets. He assumes a fan outlet pressure of about 25 mm Hg and computes the jet velocity from:

$$\bar{u}_0 = \sqrt{\frac{2gP}{(1.2)\rho_0}}$$ (90)

where ρ_0 is the density of the nozzle fluid and the factor of 1.2 compensates for 20% pressure drop through the ducting.

The required jet diameter is computed from Eq. (91)

$$d_0 = \frac{L_j}{k_1 (\bar{u}_0/u_1) \sqrt{(\rho_0/\rho_a)}} \tag{91}$$

where k_1 is 1.6 for s/d_0 = 4 to 5, and L_j is taken to be a factor times L_T, the axial length of the furnace. The factors were given on Page 157 and range from 0.8 to 1.10.

The number of nozzles in the row N is then calculated from the furnace width Z and the jet spacing s according to

$$N = \frac{Z - 4s}{s} \tag{92}$$

The required fan capacity Q_t is then computed from

$$Q_t = N \frac{\pi d_0^2}{4} \bar{u}_0 \tag{93}$$

(ii) The Bituminous Coal Research (BCR) Method. The National Coal Association has published a design handbook for Layout and Application of Overfire Jets for Smoke Control, based on work by BCR [91].

The NCA recommends:

Sidewall placement

Location of nozzles about 0.5 m above the fuel bed in modern furnaces and from 0.2 to 0.3 m above the bed in older, small furnaces

Introduction of from 10 to 30% theoretical air via jets depending on whether the smoke formed is "light" or "heavy"

The design method is as follows:

Read the required volume of air (cubic feet per minute per pound coal burned per hour) from a table, given the heating value of the coal and whether the smoke is light, moderate, or heavy. Compute the air requirement in cfm.

Decide where nozzles will be located (front, side, or back wall).

Read the number of nozzles required from a table, given the dimension of the wall on which the jets are to be located and the penetration distance (equal to the axial dimension of the furnace).

Compute the air requirement per nozzle by dividing the result of the first step and result of the third step.

Read nozzle diameter and required fan pressure from graphs, given the air requirement per nozzle (fourth step) and the penetration distance from the third step.

Determine duct size from a nomograph, given total air requirement from the first step.

The design criteria on which this method is based are not readily apparent. Examination of the tables and graphs included in the reference shows the following relationships:

The number of jets is approximately proportional to the length of the furnace wall where the jets are installed and approximately proportional to the desired penetration distance.

The penetration distance appears to be defined as that distance required to reduce the velocity of a jet, issuing into a quiescent chamber, to 2.5 m/sec.

Several qualitative statements can be made. First, the crossflow velocity does not enter explicitly into the design method. Second, working out several examples shows that relative spacing s/d_0 of up to 10 or more result. This is at odds with Ivanov's finding that spacings of four to five jet diameters are optimal.

The design methods cited above apply to furnaces burning coal or shale, and are generally used in boiler design. These applications are characterized by:

A uniform and predicatble fuel supply which burns in a regular and repeatable pattern along the grate

Use of high heating value fuel (in the range of 5500 to 8300 kcal/kg), with low moisture and ash content

The desirability of minimizing excess air so that high combustion temperatures and high heat recovery efficiency can be obtained

Relatively low combustion volume per Btu per hour capacity

In contrast, mass burning incinerators are characterized by:

A variable and generally unpredictable fuel supply; the composition and moisture content vary seasonally in a somewhat predictable manner and hourly (as fired) in an unpredictable manner

Use of low heating value fuel (2475 kcal/kg average as fired), with relatively high ash (20%) and moisture (28%) content [21]

No general requirement for high combustion gas temperature, except in heat recovering incinerators

Relatively large combustion volumes per kilocalories per hour capacity

In both cases, complete fuel burnout is desirable and combustion gas temperatures must be kept below the point where slagging or damage to the refractory occurs. Both types of units have fly ash problems, although the incinerator problem is more severe since relatively large pieces of unburned paper can be lifted into the combustion volume.

The difference in characteristics place different requirements on the overfire jets. Jet systems in mass burning incinerators must contend with

A shifting combustion profile caused by variations in the upflow gas temperature, composition, and velocity, and in the moisture content and composition of the fired refuse

Large pieces of partially burned refuse in the combustion volume

Large combustion volumes per kilocalories per hour which increases the difficulty of mixing the combustion gases

Typical excess air levels, even in incinerator boilers, upwards of 100%

In meeting these conditions, minimization of excess air introduced in the jets is not as important as in heat recovering boilers. The principal factors which mitigate for low excess air in incinerators are draft limitations, higher costs of air pollution control, fan and stack equipment, power costs, and the general requirement that the overfire air not quench the combustion reaction. Although these factors are important, realization of complete combustion of pollutants, materials survival, and inhibition of slagging are predominant concerns.

(f) Combustion Effects. The analysis in Ref. 21 and experimental data [30] indicate that under almost all conditions the flow of air through a refuse bed will produce gases containing unburned combustible. Although bypassing (channeling) of air through the bed can provide some of the needed oxidant to the fuel-rich gases, the data by Kaiser [31] indicate that the gases, on the average, remain fuel-rich. As a consequence, it should be expected that some combustion will occur as the oxygen-bearing overfire air jet penetrates the hot gas flow rising from the bed. This results in a so-called inverted flame where a jet of oxidant discharges into, "ignites," and burns in a fuel-rich environment.

This phenomenon has been observed in both coal and refuse burning practice, where impingement of a jet on the opposite sidewall has resulted in refractory overheating and slagging, contributing to premature wall failure.

In another case, a jet of air moving beneath a long arch over the discharge grate of an incinerator furnace yielded temperatures of over 2500°F in the brickwork. Because of the potential importance of this combustion effect, a simplified mathematical model of jet behavior under these conditions was developed [63], and the effect of the pertinent variables was explored.

The analysis makes use of jet concentration correlations describing the axial $\bar{c}_m$ and radial $\bar{c}$ weight concentration of nozzle fluid as functions of the distance from the nozzle plane x and the radial dimension r. The concentration at the nozzle is c_0 and the nozzle diameter is d_0. The ambient T_a, nozzle fluid T_0, and mixture T_m temperatures are those prior to combustion. For noncombusting jets, in the absence of crossflow and buoyancy, these variables are related by

$$\bar{c}_m = 5\bar{c}_0 \left(\frac{T_a}{T_0}\right)^{1/2} \left(\frac{d_0}{x}\right) \tag{94}$$

$$\bar{c} = \bar{c}_m \exp\left[-57.5\left(\frac{r}{x}\right)^2\right] \tag{95}$$

Assuming equal and constant specific heats for the nozzle and ambient fluid, an energy balance yields

$$T_m = T_a + 5\left(\frac{T_a}{T_0}\right)^{1/2} (T_0 - T_a)\frac{d_0}{x} \exp\left[-57.5\left(\frac{r}{x}\right)^2\right] \tag{96}$$

Using Eq. (96), one can calculate the mixture temperature. Then, by comparison with an assumed minimum ignition temperature T_i (say 590°C), it can be determined whether combustion will occur. (Note that this procedure takes account of "quenching" of combustion reactions in the cold core of the jet.)

The oxygen demand θ (kilograms of oxygen per kilogram of ambient fluid) and heat of combustion ΔH_c (kilocalories per kilogram of oxygen reacting) of the furnace gases can be estimated. Defining Ω as the concentration of nozzle fluid in the mixture relative to the nozzle concentration $\bar{c}/\bar{c}_0$, we find that if $T_m \geq T_i$ and if $\theta(1 - \Omega) - \Omega\bar{c}_0 \geq 0$, combustion will occur to the extent of the available oxygen releasing $\Delta H_c \Omega\bar{c}_0/\theta$ kcal/kg of mixture. The resulting gas temperature T_F is given by

$$T_F = T_R + \frac{1}{c^\circ_{p,av}}\left[(1 - \Omega)(T_1 - T_R)c^\circ_{p,av} + \frac{\Delta H_c \bar{c}_0}{\theta}\right] \tag{97}$$

where T_R is the reference temperature for enthalpy (say 15°C) and $c_{p,av}^\circ$ is the average specific heat of the gases between the reference temperature and T_F.

For the oxygen-rich case, if $T_m \geq T_i$ and if $\theta(1 - \Omega) - \Omega \bar{c}_0 < 0$ combustion will occur to the extent of the available fuel releasing $\Delta H_c(1 - \Omega)$ kcal/ kg of mixture. The resulting gas temperature is given by

$$T_F = T_R + \frac{1}{c_{p,av}^\circ} [(1 - \Omega)(T_1 - T_R)c_{p,av}^\circ + \Delta H_c(1 - \Omega)] \tag{98}$$

Calculation of the radial profiles of temperature according to the above was carried out and the results are plotted in Figs. 28, 29, and 30, at various distances from the nozzle plane. The depressed temperature (600°C) near the discharge plane is the subignition temperature or "quenched" region. The peaks of temperature found along a radial temperature "traverse" identify the stoichiometric point. The peak temperature (about 1400°C) reflects the assumptions

$$T_a = 1094°C \ (2000°F)$$

$$\Delta H_c = 206 \text{ kcal/kg } (370 \text{ Btu/lb})$$

$$\theta = 0.0545 \text{ kg } O_2/\text{kg ambient}$$

$$\bar{c}_0 = 0.23 \text{ kg } O_2/\text{kg nozzle fluid (air)}$$

$$c_{p,av}^\circ = 0.31 \text{ kcal/kg } °C^{-1} \text{ relative to } 15.5°C$$

Note that only absolute temperatures are to be substituted in the equations.

It can be seen that a hot zone is rapidly developed with a diameter of about 30 cm (1 ft) and this zone endures, for large jets, for distances which approximate the widths of many incinerator furnaces (2-3 m). This problem is greatly reduced as the jet diameter is decreased, thus adding additional incentive to the use of small diameter jets.

The rapid attainment of stoichiometric mixtures within the jet is in agreement with the theory developed by Thring [92] which roughly approximates the spread in velocity and concentration using the same mathematical formulations as in Eqs. (73) through (76), but using an effective nozzle diameter equal to $d_0(\rho_0/\rho_f)^{1/2}$ where ρ_f is the gas density at the temperature of the flame.

The above illustrates the characteristics of "blow torch" effects from air jets and shows that temperatures near 1400°C, such as have been experienced with jets, could be anticipated. Clearly, however, the quantitative accuracy of the above calculations is limited to providing only an initial estimate of the nature of air jet behavior in incinerators. No allowance is

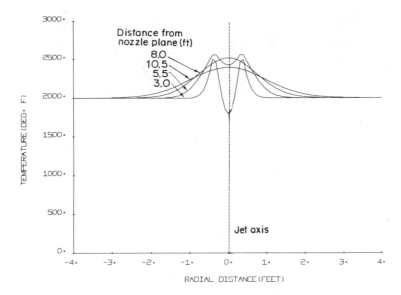

FIG. 28. Temperature versus radial distance (jet diameter = 1.5 in.) (from Ref. 63).

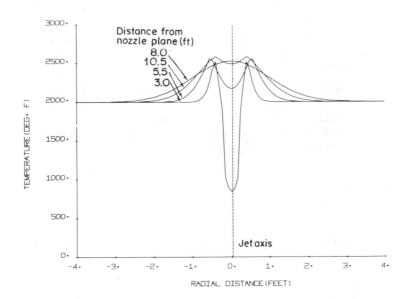

FIG. 29. Temperature versus radial distance (jet diameter = 2 in.) (from Ref. 63).

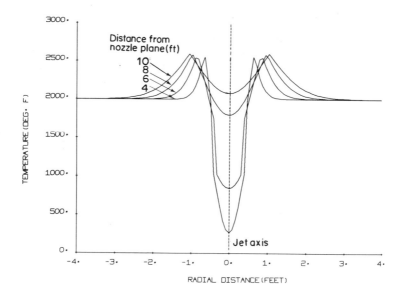

FIG. 30. Temperature versus radial distance (jet diameter = 4 in.)
(from Ref. 63).

made in the analysis, for example, of crossflow effects which are known to
increase jet entrainment rates and thus "shorten" the inverted flame described by the analysis. The method, therefore, can be expected to produce
a conservative result.

The correlations of axial concentration by Patrick [87] provide a
means to estimate the effect of crossflow in shortening the distance to the
point of completion of the combustion reactions. From Eq. (94) an analysis
readily shows the distance from the nozzle plane to the point where the gas
on the axis is at a stoichiometric ratio to be given by

$$\frac{x}{d_0} = 5 \left(\frac{T_a}{T_0}\right)^{1/2} \left(\frac{\overline{c}_0}{\theta} + 1\right) \tag{99}$$

Patrick found the centerline concentration to vary along the jet path length
ℓ in crossflow according to

$$\frac{\overline{c}_0}{\overline{c}_m} = \left[\left(\frac{\ell}{d_0}\right) \exp (7.8M^{1/2} - 1.856)\right]^{1.18} \tag{100}$$

and, for no crossflow according to

$$\frac{\overline{c}_0}{\overline{c}_m} = 0.112\left(\frac{x}{d_0}\right)^{1.18}$$ (101)

Combining Eqs. (100) and (101) establishes the relationship between the centerline distance x for noncrossflow which corresponds to the same concentration ratio as for a jet in crossflow which has traveled over a path length ℓ

$$\frac{x}{d_0} = 1.42\left(\frac{\ell}{d_0}\right) \exp\,(7.8M^{1/2} - 1.856)$$ (102)

The path length can be easily calculated by numerical integration of Eq. (103) which also can be derived from Patrick's trajectory relationships:

$$\left(\frac{\ell}{d_0}\right) = \int_0^{x/d_0} \left[9\,M^{2.55}\left(\frac{x}{d_0}\right)^4 + 1\right]^{0.5} d\left(\frac{x}{d_0}\right)$$ (103)

Therefore, to find the distance from the nozzle plane to the point where such peak (stoichiometric) temperatures will be obtained on the centerline, the noncrossflow distance is calculated from Eq. (99); the resulting value is substituted into Eq. (102) to yield the crossflow path length at an equivalent degree of mixing; and the integration given in Eq. (103) is carried out to define the dimensionless distance x/d_0 integration limit which causes the integral to assume the value of the calculated path length. This latter x/d_0 value corresponds to the horizontal distance from the nozzle to the plane where peak temperatures exist. The vertical displacement of the jet in this plane may then be calculated by substitution into Eq. (104).

$$\left(\frac{y}{d_0}\right) = M^{1.28}\left(\frac{x}{d_0}\right)^3$$ (104)

The analysis shows that jet temperatures can be considerably elevated by combustion effects. Therefore, when analyzing jet operation in regimes where a buoyancy analysis (neglecting combustion) suggests jet drop will be important, these effects could provide counterbalancing jet temperature increases.

(g) Incinerator Overfire Air Jet Design Method. The basic parameters to be selected in design of an overfire air jet system are:

The diameter (d_0) and number (N) of the jets to be used

The placement of the jets

The quantity of the air to be overfired

Related but not independent variables are the jet velocity $\bar{u}_0$ and the head requirements for the overfire air fan P.

The tentative design method is based on that of Ivanov [81] which was discussed above. It is important in using this method to substitute the values of d_0 and u_1 obtained into Eq. (88) to determine if buoyant forces might be important. Values should fall in the "drag forces predominate" region.

The basic equations on which the design method is based are as follows:

Air Flow Relation

$$Q_T = 47 \, N d_0^2 \bar{u}_0 \tag{105}$$

where

Q_T = the overfire air rate (m^3/min)

N = the number of jets

d_0 = the jet diameter (m)

$\bar{u}_0$ = the jet velocity (m/sec)

Ivanov's Penetration Equation

$$\frac{L_j}{d_0} = 1.6 \left(\frac{\bar{u}_0}{u_1} \right) \sqrt{\frac{\rho_0}{\rho_a}} \tag{106}$$

where

L_j = the desired jet penetration

u_1 = the estimated crossflow velocity in the incinerator

ρ_0 and ρ_a = the jet and crossflow gas densities, respectively

Jet Spacing Equation

$$z = \left(\frac{s}{d_0} \right) N d_0 \tag{107}$$

where

z = the length of furnace wall on which the jets are to be placed

$\frac{s}{d_0}$ = the desired value of jet spacing (measured in jet diameters)

Inherent in Eq. (107) is the assumption that the jets are placed in a single line. Ivanov recommends that s/d_0 be in the range of 4 to 5, although values as low as 3 can probably be used without invalidating the penetration equation [Eq. (106)].

The form of these equations sheds some light on the options open to the systems designer. For a given set of furnace conditions, Eq. (106) can be rearranged to give

$$d_0 \bar{u}_0 = \text{Constant} \tag{108}$$

The product of jet diameter and velocity is fixed by furnace conditions. Substitution of Eqs. (107) and (108) into Eq. (105) yields

$$\frac{Q_T}{z} = \frac{0.326 d_0 \bar{u}_0}{s/d_0} \tag{109}$$

The air flow per length of wall is fixed by furnace conditions, except inasmuch as s/d_0 can vary from 3 to 5.

Example 14. Consider an incinerator with the following characteristics

Capacity, 250 metric tons per day (24 hr basis)

Total stoichiometric air requirement, 425 m^3/min

Length of wall for jet placement L, 9 m

Desired depth of penetration L_j, 2.5 m

Crossflow velocity u_1, 1.25 m/sec

Jet and furnace temperatures of 40°C (313 K) and 1090°C (1363 K), respectively, so that

$$\sqrt{\frac{\rho_a}{\rho_0}} = \sqrt{\frac{T_0}{T_a}} = \sqrt{\frac{313}{1363}} = 0.48$$

Develop the design of the overfire jet system.

<u>Step 1.</u> Compute the product $d_0 \bar{u}_0$ from Eq. (106)

$$d_0 \bar{u}_0 = L_j u_1 \left(\frac{1}{1.6}\right) \sqrt{\frac{\rho_a}{\rho_0}} \tag{110}$$

$$= \frac{(2.5)(1.25)(0.48)}{1.6} = 0.94 \text{ m}^2/\text{sec}$$

<u>Step 2.</u> Select value of $\bar{u}_0$. Using Ivanov's guidelines, set $\bar{u}_0 = 60$ m/sec (cold jet).

<u>Step 3.</u> Compute d_0 from the result of Step 1.

$$d_0 = \frac{0.94}{60} \text{ m} = 0.0157 \text{ m} = 1.57 \text{ cm}$$

The pressure requirement for the overfire air fan depends mainly on $\bar{u}_0$. Eq. (103) may be used to calculate the required velocity head as a function of $\bar{u}_0$.

$$\Delta P = 1.2 \left[\frac{\rho_0 \bar{u}_0^2}{2g}\right] \tag{111}$$

for air at 40°C, $\rho_0 = 1.12$ kg/m^3, and for $\bar{u}_0 = 60$ m/sec, $\Delta P = 247$ kg/m^2 or 46.2 cm Hg.

<u>Step 4.</u> Compute the number of jets N. Setting $s/d_0 = 4$, Eq. (107) yields

$$N = \frac{9}{(4)(0.0157)} = 143 \text{ jets}$$

<u>Step 5.</u> Compute Q_T from Eq. (105).

$$Q_T = 47(143)(0.157)^2(60) = 99.4 \text{ m}^3/\text{min}$$

<u>Step 6.</u> Compare Q_T with the theoretical air requirement. The theoretical air requirement is 425 m^3/min, so that the overfire air rate is equivalent to $(99.4/425)(100) = 23\%$ of theoretical. The soundness of this value can be checked by comparison with the air requirements defined by the bed burning process and by reference to experience. It is worthy to note, however, that few data exist to allow confident valuation of the performance of existing plants with respect to combustible pollutant emissions and the design and operating parameters of the overfire air systems.

The amount of overfire air can be increased or decreased within limits without seriously affecting the performance of the jets by changing the jet spacing parameter [s/d_0 in Eq. (107)] within the range of 3 to 5. It should be recognized that the design method described has as its goal the complete mixing of fuel vapors arising from the fuel bed with sufficient air to complete combustion. Two other criteria may lead to the addition of supplementary air quantities: the provision of sufficient secondary air to meet the combustion air requirement of the fuel vapors, and the provision of sufficient air to temper furnace temperatures to avoid slagging (especially in refractory-lined incinerators). In the example given, both criteria act to greatly increase the air requirement and would suggest the installation of additional jets (although the discharge velocity criteria are relaxed). Nonetheless, the Ivanov method is intended to assure the effective utilization of overfire air and, in boiler-type incinerators where tempering is not required, will permit optimal design and operating efficiency.

(h) Isothermal Slot Jets. The discussions above have dealt with the round jet, the configuration found almost exclusively in combustion systems. In designs where a multiplicity of closely spaced round jets are used, however, the flow fields of adjacent jets merge. The flow then takes on many of the characteristics of a jet formed by a long slot of width y_0.

For use in analyzing such situations, the flow equations applicable to such a slot jet (or plane jet) are:

Centerline velocity

$$\frac{\overline{u}_m}{\overline{u}_0} = 2.48 \left(\frac{x}{y_0} + 0.6 \right)^{-1/2} \left(\frac{\rho_0}{\rho_a} \right)^{1/2} \tag{112}$$

Centerline concentration

$$\frac{\overline{c}_m}{\overline{c}_0} = 2.00 \left(\frac{x}{y_0} + 0.6 \right)^{-1/2} \left(\frac{\rho_0}{\rho_a} \right) \tag{113}$$

Transverse velocity

$$\frac{\overline{u}}{\overline{u}_m} = \exp \left[-75 \left(\frac{y}{x} \right)^2 \right] \tag{114a}$$

$$= 0.5 \left[1 + \cos \left(\frac{\pi y}{0.192x} \right) \right] \tag{114b}$$

Transverse concentration

$$\frac{\bar{c}}{\bar{u}_m} = \exp\left[-36.6\left(\frac{y}{x}\right)^2\right]$$
(115a)

$$= 0.5\left[1 + \cos\left(\frac{\pi y}{0.279x}\right)\right]$$
(115b)

These functions indicate a velocity half-angle of 5.5° and a concentration half-angle of 7.9°.

Entrainment

$$\frac{\dot{m}_x}{\dot{m}_0} = 0.508\left(\frac{\rho_a}{\rho_0}\right)^{1/2}\left(\frac{x}{y_0}\right)^{1/2}$$
(116)

where $\dot{m}_x$ and $\dot{m}_0$ are for a unit length of slot and $\dot{m}_0 = \rho_0\bar{u}_0 y_0$

b. Swirling Flows

In waste and fuel burners and within furnaces, swirl is often used to modify flow characteristics. Because of the intense recirculation patterns in swirling flows (burning gases travel back towards the burner bringing heat energy and reactive species to promote ignition in the entering fuel-air mixture), rotation is found to shorten the flame. The recirculation and flame shortening effects may be exploited to reduce the size and cost of the enclosure or to increase the post-flame residence time.

(1) The Axial Swirl Burner—Isothermal Performance

Velocities in axially symmetrical swirling jets (the axial swirl burner) can be defined in terms of three components. Axial velocity $\bar{u}$ is the component parallel to (but not necessarily on) the jet axis; radial velocity $\bar{v}$ is the velocity towards or away from the axis; and $\bar{w}$ is the velocity tangent to a circle centered on the axis.

The intensity of swirl in a jet has been characterized using the swirl number N_s which is defined as the dimensionless ratio of jet angular momentum $\dot{L}_0$ to linear momentum $\dot{G}_0$:

$$\dot{L}_0 = 2\pi\rho_0\int_0^{r_0}(\bar{u}\bar{w})_0 r^2 dr$$
(117)

$$\dot{G}_0 = 2\pi \int_0^{r_0} (P + \rho \bar{u}^2)_0 r dr \tag{118}$$

$$N_s = \frac{2\dot{L}_0}{(\dot{G}_0 d_0)} \tag{119}$$

Where the static pressure P (gauge) in Eq. (118) allows for pressure variations over the nozzle cross-section owing to centrifugal forces. Swirl numbers of typical burners are usually in the range 0.6 to 2.5 [93].

The swirl burner can be constructed in a number of different configurations including straight or profiled vanes, tangential entry with radial vanes, a movable block system [94], or a scroll-type tangential inlet. These systems, though geometrically different, lead to highly similar flow patterns.

A large, toroidal recirculation zone is formed in the exit, occupying up to 75% of the exit diameter, with up to 80% of the initial flow being recirculated.

Very high levels of kinetic energy of turbulence (KE = $(\overline{u'^2} + \overline{w'^2} + \overline{v'^2}/u_0^2)$) are formed (up to 300% of the kinetic energy at the nozzle plane) with a rapid decay (to less than 50%) within one exit diameter.

The swirling flow exhibits a three-dimensional, time dependent instability called the precessing vortex core (PVC) which can lead to the generation of tonal combustion noise or lead to blow-out and other instabilities [93].

The axial reverse flow (initiating at $N_s > 0.6$) is, surprisingly, virtually nonswirling and has a magnitude M_r relative to the nozzle flow (for straight exits) given by:

$$N_s = 0.508 + 5.66M_r - 6.25M_r^2 + 2.28M_r^3 \tag{120}$$

The presence of an oil gun or other source of nonswirling axial flow can reduce M_r by up to 50%, while a divergent exit nozzle can increase M_r by 500 to 600% over that for straight exits, with M_r increasing with increasing divergence angle (say, 24° to 35° half-angle) up to an as yet undefined limit where M_r ultimately begins decreasing (as detachment occurs).

The shape of the reverse flow zone is dependent upon the type of swirler with more compact, smaller volume zones being formed with tangential inlet swirlers and longer, thinner zones being formed with vane swirlers. Thus the tangential inlet swirlers have higher reverse flow velocities and steeper velocity gradients (more intense turbulence and consequent mixing), since the M_r for both swirlers and comparable.

The rate of entrainment of ambient fluid and jet half-angle is much higher for swirling flows than that in nonswirling jets. Experimental data by Kerr and Fraser [95] indicate approximately linear entrainment and half-angle laws:

$$\frac{\dot{m}_x}{\dot{m}_0} = (0.35 + 0.7N_s)\frac{x}{d_0} \tag{121}$$

$$\tan \phi_s = (1 + 2.84N_s)\tan \phi_0 \tag{122}$$

where

$\dot{m}_x, \dot{m}_0$ = the mass flows at a distance x and at the plane of the nozzle of diameter d_0, respectively

ϕ_s, ϕ_0 = the jet half-angle with and without swirl, respectively (see Section C.1.a(3a).

Especially at high N_s and when the ratio of the cross-sectional dimension of the burner d_0 to that of the enclosure d_c is high, a second recirculation zone is established at the wall. The mass flow of this outer recirculation may be greater or less than the central zone. Although the central zone is of greater interest to combustion engineers, the existence of the wall flow can be of great influence on wall heat transfer rates and consequent materials problems.

The relationship between pressure drop and N_s is dependent upon swirler design, although, in all cases pressure loss coefficient $\Delta P/(1/2)\rho_0 u_0^2)$ increases with increasing N_s. Annular vaned swirlers are particularly sensitive to N_s variations, with the pressure loss coefficient (PLC) increasing from 8 to 50 as N_s increases from 1 to 2.8. Tangential entry swirlers showed a three-fold increase in PLC as the Reynolds number (based on swirler diameter) increased from 0.3×10^5 to 2.3×10^5. Studies on the effect of an axial oil gun [96] showed a three-fold increase in PLC if a gun was used. Considerable reduction of PLC was shown for small values of d_0/d_c. Typical ranges of d_0/d_c lie between 0.4 and 0.7 [96]. A divergent outlet also appears to reduce the PLC, typically by 10 to 20%.

(2) The Cyclone Combustion Chamber—Isothermal Performance

A second type of swirling combustor is the cyclone combustion chamber, where air and fuel are injected tangentially into a large usually cylindrical chamber and exhausted through a centrally located exit hole in one end. These combustors are often used for the combustion of materials which are normally considered difficult to burn efficiently, such as damp vegetable refuse, high ash content, and brown coals, anthracite, and high sulfur oils.

There are two main types of such chambers. The first (Type A) has two or more tangential inlets into the top of a cylindrical chamber, with exit through a constricted orifice in the bottom, and is commonly used for higher calorific value fuels where slag and ash generation and removal is not a serious problem. The second (Type B) has one to four tangential inlets at the top, a baffled exit at the top, and a slag tap at the bottom, and is used with high ash content fuels.

The swirl number for such systems, assuming a uniform exit axial velocity profile, can be related to dimensional parameters:

$$N_{sc} = \frac{\pi d_e d_c}{4A_t} \tag{123}$$

where d_e is the diameter of the diameter of the exit throat and A_t is the area of the tangential inlet.

On this basis, Type A systems usually operate at an N_{sc} from 2 to 11, a length to exit diameter L_T/d_e of from 1 to 3, and a ratio of d_e/d_c of from 0.4 to 0.7. Type B systems operate at an N_{sc} from 8 to 20, a L_T/d_e from 1 to 1.25, and a d_e/d_c from 0.4 to 0.5. At these N_{sc}, very high residence times are achieved, and most of the combustion processes appear to occur inside the chamber.

The flows within Type A and Type B chambers are complex and sensitive to the inlet and outlet configurations, relative inlet, outlet, and chamber dimensions, and to the presence of vanes or other flow modifiers on the sidewalls or endplates [93].

The pressure drop through both types of cyclones has been expressed by Tager [97] as the sum of inlet losses (ΔP_i) and losses across the chamber (ΔP_{cc}):

$$\Delta P = \Delta P_i + \Delta P_{cc} \tag{124}$$

$$\Delta P_i = \xi_a \frac{\overline{w}_{in}^2 \rho_0}{2g} \tag{125a}$$

$$\Delta P_{cc} = \xi_{cc} \frac{\overline{u}_e^2 \rho_0}{2g} \tag{125b}$$

where

$\overline{w}_{in}$ = the tangential inlet velocity

$\overline{u}_e$ = the average exit velocity

Values of the coefficients ξ_a and ξ_{cc} can be taken from Fig. 31 for d_e/d_c of 0.45. Tager's correlation to other values of this parameter [97] showed for d_e/d_c from 0.3 to 1.0:

$$\Delta P_{cc} = (\Delta P_{cc})_{0.45} \left(\frac{0.9}{d_e/d_c} - 1.0 \right) \tag{126}$$

The efficiency of swirl generation in cyclone combustors is usually in the range 4 to 15%. This is to be compared with efficiencies as high as 70 to 80% in swirl burners. For Type A cyclones, Fig. 32 shows the effect of variation in constructional parameters on efficiency [98].

(3) The Axial Swirl Burner—Combustion Effects

Combustion processes (especially volumetric expansion) exert a great influence on swirling flows. Few data are available to quantify the impact

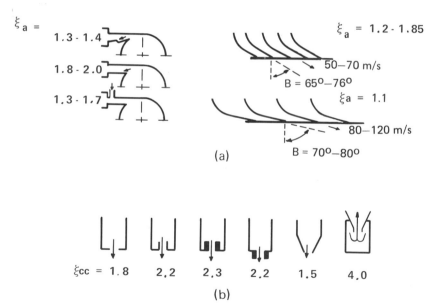

(a)

(b)

FIG. 31. (a) Inlet losses for various types of swirlers. (b) Chamber-friction, swirl, and outlet losses. (From Ref. 97.)

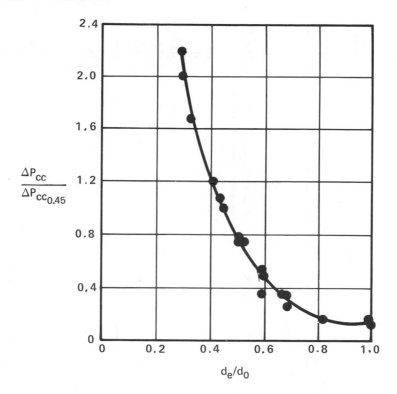

FIG. 32. Dependence of the aerodynamic efficiency on the constructional parameters of Type A cyclone combustors. Points obtained from
tests on eight different cyclones in self-similarity Reynolds number range
(Re > 50,000) by Tager (from Ref. 97).

of combustion (especially ϕ, the mixture ratio relative to stoichiometric).
An example for unenclosed swirl burners is presented in Table 23.

It is apparent, however, from these and other data [99,100] that with
combustion in comparison to the isothermal case:

A considerable reduction in M_r occurs, particularly near $\phi = 1$
and with premix.

The maximum axial velocity is almost twice that in the isothermal
case, although the decay of velocity is similar.

The density changes from combustion cause the initial jet spread
to be greater.

TABLE 23

Characteristics of Unenclosed Swirl Burners[a]

N_S	ϕ[b]	Combusting M_r	Isothermal M_r	Length of recirculating zone[c]		Width of recirculating zone[c]		Type of fuel entry	Type of swirler
				Combusting	Isothermal	Combusting	Isothermal		
0.7	1.63	0.005	0.04	0 to 2 d_e	3.5 d_e	0.83 d_e	0.4 d_e	Premix	Annular vaned
1.25	1.63	0.06	0.16	0 to 3 d_e	5 d_e	1.8 d_e	0.6 d_e	Premix	Annular vaned
2.2	13.8	0.64	1.00	$-1d_e > L > 2d_e$	~5 d_e	0.6 d_e	~0.7 d_e	Axial	Tangential

[a] From Ref. 93.
[b] Town gas ($\approx$ 50% H_2) was used for all of these results.
[c] Distances expressed in exit throat diameters.

The recirculation zones are shorter and wider for premixed burners. With burner mixing, the mode of fuel entry and ϕ can lead to either narrower or wider zones. Limited data [93] suggest that very high values of ϕ or axial/radial fuel entry produces zone widths less than the isothermal case.

The effect of enclosure upon a swirl burner is, as in the isothermal case, dependent upon the ratio of furnace d_c to burner d_e diameter. Although the various interactions are complex functions of ϕ, mode of fuel entry, and geometry, one can determine that the impingement point of the flame on the wall lies between $0.2d_c$ and $0.5d_c$ in the range:

$$0.7 < N_s < 1.3 \quad 3 < \frac{d_c}{d_e} < 5 \quad 0.9 < \phi < 1.7$$

The swirl parameter N_s tends to decrease due to combustion which increases the axial momentum flux. Beltagui and Maccallum [100] have suggested a 10% decrease may be appropriate.

Pressure drop increases to between 125% and 300% of the isothermal case, the larger values being obtained in premixed systems.

In understanding the behavior of swirl burner combustion the concept of a well-stirred zone followed by a plug flow zone has proven to be a powerful concept. The well-stirred zone, brought about by intense backmixing and turbulence, is of uniform composition and temperature. The burnedness β, the degree to which combustion is completed, in the well-stirred zone is generally less than unity. Thus the plug flow zone is important in providing the time for combustion to be completed.

Similarity between flows in hydraulic (water) models and actual furnaces has been studied by Beér et al. [101] and their results are shown in Figs. 33 and 34. Note that $\bar{t}$ is the mean residence time, t is the actual time following cessation of tracer injection, and $\bar{t}_s$ is the residence time in the well-stirred section. Further, N_s in the combustion cases has to be multiplied by the density ratio $(\rho_f/\rho_{iso})^{1/2}$ to compensate for effect of combustion.

Figure 33 shows the decay rates of injected tracers in a pulverized coal furnace and in a 0.1 scale water model. It can be seen that the effect of swirl is to lengthen the time until decay of tracer concentration commences and then to result in a more rapid concentration decay rate than the nonswirl case.

Figure 34 shows that the residence time distribution between the stirred zone and the plug flow zone may be controlled by swirl, going through a minimum as N_s increases. Some data [102] have shown that for oil fired combustion, operation at the minimum point in well-stirred residence time yields maximum performance with minimum smoke emissions.

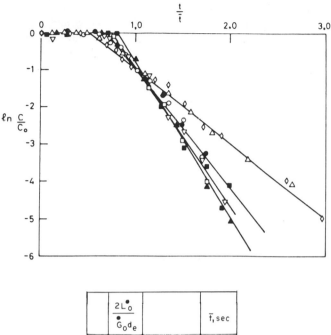

	$\dfrac{2L_0^\bullet}{G_0 d_e}$		$\bar{t}$, sec
◊	0	FURNACE	15.2
△	0	MODEL	20
■	1.56	FURNACE	18.3
▲	1.56	MODEL	20
□	2.24	FURNACE	17.9
▽	2.24	MODEL	20
○	3.9	FURNACE	19.3
●	3.9	MODEL	20

FIG. 33. Tracer concentration decay as a function of swirl number in model and prototype furnace (from Ref. 101).

(4) The Cyclone Combustion Chamber—Combustion Effects

Data on cyclone systems with combustion are few; due, importantly, to the experimental difficulties encountered with the high ash or slagging fuels most often fired in this type of equipment. Some conclusions, however, can be drawn:

Optimum combustion can be obtained by admitting 2 to 3% of the combustion air on the axis. Axial introduction of fuel leads to incomplete mixing and poor combustion.

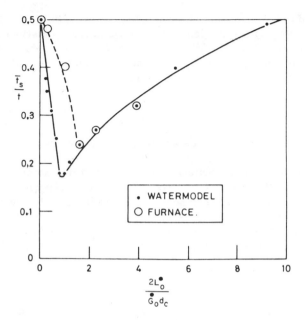

FIG. 34. Ratio of residence time in stirred section to total residence time as a function of swirl number (from Ref. 101).

At least two symmetrically arranged tangential inlets should be used to avoid uneven burning and excessive pressure drop.

In combustion situations, up to three major zones of oxygen deficiency occur, one commonly occurring in the region of the exit throat. This phenomenon, perhaps due in part to centrifugal stratification effects, leads to the need for an afterburning chamber to obtain complete burnout.

Comparisons between isothermal and combusting system regarding flow pattern and pressure drop show strong similarity. For evaluating N_S in the combusting state, the following correction is recommended:

$$(N_S)_{combusting} = (N_S)_{isothermal} \left[\frac{T \text{ inlet } °K}{T \text{ outlet } °K} \right] \tag{127}$$

2. Induced Flow

For the combustion systems and subsystems described above, emphasis was placed on the flow elements directly within the control of the designer, the axial and sidewall jet and the swirled jet or cyclonic flow. This section explores the consequential effects of these flows (recirculation) and the impact of buoyancy forces on furnace flow. Each of these induced flows has a significant effect on combustor performance and should be understood by both the designer and the trouble-shooter.

a. Jet Recirculation

As a consequence of viscous and (at higher Reynolds numbers) turbulent momentum transfer, gradients in velocity always produce an acceleration of the slower flow and proportional deceleration of the faster flow. This process leads to the generation of recirculation patterns within furnaces which can, in several circumstances, cause problems. It is not always vital that the designer be able to quantify these effects, but an awareness of their existence and potential impact is of use in anticipating and diagnosing problems. Since sidewall and axial jets are commonplace in combustors, their behavior merits special attention.

(1) Sidewall Jets

Recirculation patterns are generated within furnaces with sidewall jets by the entrainment of furnace gases into the jet flow. The entrainment coefficient J, defined as the ratio of the influx velocity component perpendicular to the jet axis $\bar{v}_m$ to the centerline jet velocity $\bar{u}_m$, can be estimated for round jets as follows:

Identify the jet's outer surface as that bounded by the points where the velocity is one-half of the centerline velocity. This surface is conical and may be characterized by the jet half-angle ϕ_0. The jet diameter is then $2x \tan \phi_0$, and the area of influx over a distance dx is $2\pi x \tan \phi_0 \, dx$.

The influx mass rate is $\dot{m}_x$, given by Eq. (77) where $\dot{m}_0 = 1/4 \bar{u}_0 \rho_0 \pi d_0^2$. The influx volume flow is $\dot{m}_x/\rho_x$. Then, the influx velocity and the entrainment coefficient are given by:

$$\bar{v}_m = \frac{1}{2\pi\rho_a x \tan \phi_0}\left(\frac{d\dot{m}_x}{dx}\right) \tag{128}$$

$$J = \frac{\bar{v}_m}{\bar{u}_m} = 0.0063 \, \mathrm{Ctn} \, \phi_0$$

and, for

$$\phi_0 = 4.85° \qquad J = 0.075$$

and   (130)

$$\bar{v}_m = 0.47 \bar{u}_0 \left(\frac{\rho_0}{\rho_a}\right)^{1/2} \frac{d_0}{x}$$

In many instances, a linear array of round jets are used. For many small jets, positioned close together, the jet flows merge a short distance from the nozzle and the flow field becomes similar to that of a long slot jet.

For an array of round jets of diameter d_0 and for a distance s center to center, the width y_0 of a slot jet of equivalent mass flow is given by

$$(y_0)_{equivalent} = \frac{\pi d_0^2}{4s} \tag{131}$$

For this situation or for the slot jet itself:

$$J = 0.051 \tag{132}$$

$$\bar{v}_m = 0.127\bar{u}_0\left(\frac{y_0}{x}\right)^{1/2}\left(\frac{\rho_0}{\rho_a}\right)^{1/2} \tag{133}$$

Example 15. A round jet with a diameter of 10 cm is situated on the side-wall of a furnace. Air at 27°C is discharged into the furnace at a velocity of 15 m/sec. Calculate the influx velocity as the jet moves across the furnace. Do the same for a linear array of 1.5 cm diameter jets on 6 cm centers with a $\bar{u}_0$ of 60 m/sec.

The mean temperature of the furnace gases is 1027°C and has a mean molecular weight approximately the same as air.

For the round jet, noting that for similar molecular weights the density ratio is given by the inverse of the absolute temperature ratio

$$\bar{v}_m = 0.47\bar{u}_0\left(\frac{\rho_0}{\rho_a}\right)^{1/2}\frac{d_0}{x}$$

$$= 0.47(15)\left(\frac{1300}{300}\right)^{1/2}(0.1)\left(\frac{1}{x}\right)$$

$$= \frac{1.47}{x}$$

For the array of jets, the equivalent slot width is given by:

$$(y_0)_{equivalent} = \frac{\pi d_0^2}{4s}$$

$$= \frac{\pi(0.015)^2}{4(0.06)}$$

$$= 0.00295 \text{ m}$$

$$= 0.295 \text{ cm}$$

$$\overline{v}_m = 0.127\overline{u}_0 \left(\frac{y_0}{x}\right)^{1/2} \left(\frac{\rho_0}{\rho_a}\right)^{1/2}$$

$$= 0.127(60)(0.00295)^{1/2} \left(\frac{1300}{300}\right)^{1/2} \left(\frac{1}{x}\right)^{1/2}$$

$$= 0.861\left(\frac{1}{x}\right)^{1/2}$$

The resulting velocity field is shown in Fig. 35.

The primary significance of these calculations lies in the demonstration that, in the region of the nozzle face, a substantial velocity field will exist, with furnace gas moving along the wall toward the jet. This flow increases the convective heat flux to the wall in the vicinity of the nozzle (raising wall temperature) and scours the wall with entrained fly ash.

If the resulting wall temperature is below the fly-ash sticking temperature, some (though probably minor) wall erosion may occur. If the wall temperature is high enough that the fly ash adheres, slag build-up will occur. This phenomenon is well known in municipal refuse incinerators where slag accumulations surround the jet orifice. Slumping or melting of the slag can occur until the overhanging mass is chilled by the cold air jet. The resulting overhang of slag can deflect the jet into the refuse bed, increasing the entrainment of fly ash and, if severe deflection occurs, resulting in a blowpipe heating effect with consequent destruction of the metal grate.

Beyond the specific effects of jet driven recirculation noted above, it is well to keep in mind the more general concept that regions of steep velocity gradients will always lead to an inspiration effect with consequent perturbation of the flow field.

(2) Axial Jets

The discharge of a jet of fuel and/or air down the axis of a combustor induces recirculation flow which, in extreme cases, can be the controlling influence in establishing combustor performance. Such cases include the long, cylindrical axially fired liquid waste incinerator and most rotary kilns.

The theory of Thring and Newby [103], though supplanted in refinement and accuracy by the work of Craya and Curtet [104], Barchilon and Curtet [105], and Hill [106], provides for relatively simple formulations to identify the approximately location of the point C where disentrainment begins (Fig. 36a).

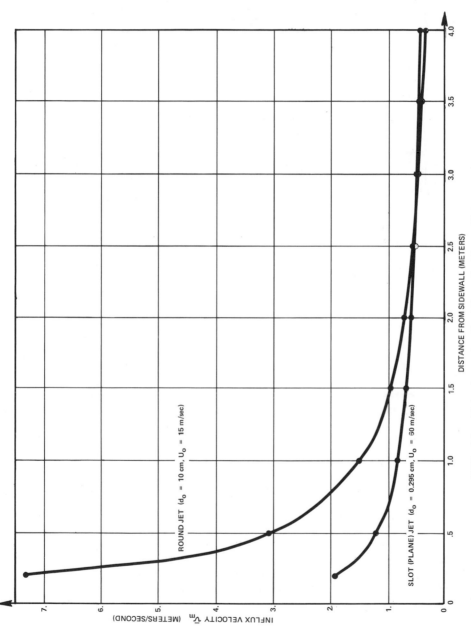

FIG. 35. Entrainment velocities of side-wall jets.

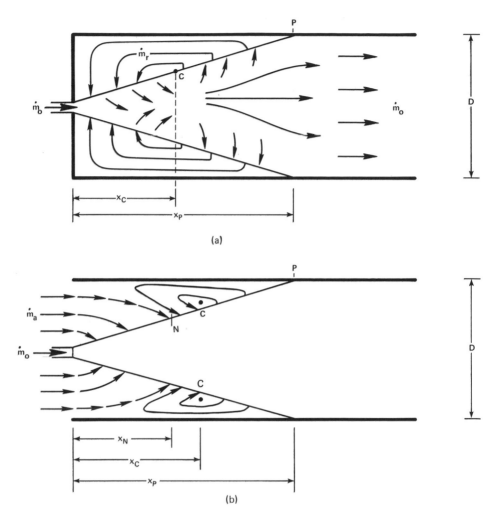

FIG. 36. Recirculation flows in cylindrical chamber: (a) flow in chambers with one end closed; (b) flow in open-ended chamber.

The Thring-Newby theory (here, modified by Field et al. [13] assumes that the jet entrains as a free jet up to the point C (at $x = x_C$). Thus, the total entrainment up to this point (the net recirculation flow $\dot{m}_r$) is given, using Eq. (77) as

$$\dot{m}_r = \left[0.32 \left(\frac{\rho_a}{\rho_0} \right)^{1/2} \frac{x_C}{d_0} \dot{m}_0 \right] - \dot{m}_0 \tag{134}$$

The point C is postulated to lie midway between the point x_0 where entrainment begins $(\dot{m}_x = \dot{m}_0)$, according to Eq. (77) and the point P where the envelope of an unconfined jet would strike the wall. From Eq. (77):

$$x_0 = \frac{d_0}{0.32}\left(\frac{\rho_0}{\rho_a}\right)^{1/2} \tag{135}$$

Assuming a jet angle of 9.7°, the jet strikes the wall at a distance x_P which, relative to the duct diameter D, is given by:

$$x_P = 2.925D \tag{136}$$

Thus, x_C is given by

$$x_C = \frac{1}{2}\left[x_P + \frac{d_0}{0.32}\left(\frac{\rho_0}{\rho_a}\right)^{1/2}\right]$$

A parameter ω is then defined:

$$\omega = \frac{d_0}{D}\left(\frac{\rho_0}{\rho_a}\right)^{1/2} \tag{138}$$

hence

$$x_C = \frac{1}{2}\left(x_P + \frac{D\omega}{0.32}\right)$$

$$= D\left(1.467 + \frac{\omega}{0.64}\right)$$

$$\frac{\dot{m}_r}{\dot{m}_0} = \frac{0.47}{\omega} - 0.5$$

For the case where a secondary air flow of mass flow $\dot{m}_a$ but of negligible momentum is introduced (as in a kiln), it is assumed that the secondary air is entrained by the primary jet flow before any recirculating gas is entrained. For this situation, an equivalent nozzle diameter d_0' is calculated as

$$d_0' = \frac{2(\dot{m}_0 + \dot{m}_a)}{(\pi\rho_0\dot{m}_0\bar{u}_0)^{1/2}} \tag{140}$$

Equation (139) still holds, but now the left side is $\dot{m}_r/(\dot{m}_0 + \dot{m}_r)$ and ω' is given by

$$\omega' = \left(\frac{\dot{m}_a + \dot{m}_0}{\dot{m}_0}\right) \frac{d_0'}{D} \left(\frac{\rho_0}{\rho_a}\right)^{1/2} \tag{141}$$

The theory then gives the position N, where recirculation entrainment begins (x_N) and the position C, which is the core of the recirculation eddy (x_C) as

$$x_N = 3.125\,\omega'D \tag{142}$$

and

$$x_C = 1.56(\omega' + 0.94)D \tag{143}$$

The distance to the point P where the jet strikes the wall (x_P) is as in Eq. (134) and

$$\frac{\dot{m}_r}{(\dot{m}_0 + \dot{m}_a)} = \frac{0.72}{\omega'} - 1 \tag{144}$$

Data from cold models and hot furnaces show that Eq. (139) fits the slope of the data but is consistently high in its estimation by an increment of about 0.25 in a plot of $\dot{m}_r/\dot{m}_0$ vs $1/\omega$. A similar comparison of data using Eq. (144) (plotting vs. $1/\omega'$) overpredicts the mass flow ratio by up to a factor of 1.7 (at $1/\omega' = 7$) but joins the data at $1/\omega' = 2$.

The position of points N, C, and P, in data compared by Field et al. [13] gave a reasonable fit. The fit was improved, however, by using a narrower effective jet angle (9.2° vs. 9.7°) in estimating x_P.

It should be noted that these analytical expressions are valid only for d_0'/D less than about 0.05.

b. Buoyancy

In many incinerator furnaces, it is desired to minimize gas velocities. This design objective is usually established to minimize entrainment of particulate matter. When boiler tubes are immersed in the flow, avoidance of erosion metal wastage is also important.

One approach to estimating the velocities in the furnace imagines all heat to remain in the gas (thus giving the gases a maximum volumetric expansion), and the to calculate the mean velocity by dividing the volumetric flow rate by the furnace cross-sectional area. Although such an approach may be useful for preliminary estimates, the potential impact of buoyancy forces, accelerating the gases to many times the "average" velocity, should be carefully considered.

The acceleration of gases from an initial velocity u_0, elevation y_0, and pressure P_0 is described by Bernoulli's Equation:

$$u^2 = u_0^2 + \frac{2(P_0 - P)}{\rho_0} g_c - 2g(y - y_0) \tag{145}$$

<u>Case 1.</u> The furnace may be considered to be well sealed and to consist of two zones: (1) a hot gas flow, and (2) a cold, stagnant (or slowly moving) gas zone

This case might be typified by a sealed municipal incinerator with the hot gases arising from the refuse bed and the colder gases (over the residue quench tank) moving out slowly. Here, the change in static pressure $(P_0 - P)$ for the cold gas, as the flows rise through a vertical outlet flue, is also experienced by the hot gas. Noting that the ratio of absolute temperatures is the inverse of the ratio of densities, writing Eq. (145) for both flows and combining yields

$$(u^2)_{hot} = (u_0^2)_{hot} + 2g(y - y_0)\left(\frac{T_{hot}}{T_{cold}} - 1\right) \tag{146}$$

<u>Case 2.</u> As for Case 1 but with a "leaky" furnace so there is ready communion with the atmosphere

This case is common for many older municipal incinerators. Under these conditions, there is an interaction between the hot and cold gases and ambient with the both furnace gas streams accelerating, but with the acceleration of the hot zone being more pronounced. In this case

$$(u^2)_{hot} = (u_0^2)_{hot} + 2g(y - y_0)_{hot}\left(\frac{T_{hot}}{T_a} - 1\right) \tag{147a}$$

$$(u^2)_{cold} = (u_0^2)_{cold} + 2g(y - y_0)_{cold}\left(\frac{T_{cold}}{T_a} - 1\right) \tag{147b}$$

and for, roughly, equal elevation changes   (147c)

$$(u^2)_{hot} - (u^2)_{cold} = (u_0^2)_{hot} - (u_0^2)_{cold} + 2g(y - y_0)_{av}\left(\frac{T_{hot} - T_{cold}}{T_a}\right)$$

Example 16. In a large, well-sealed furnace, 6000 m^3/min of gases leave
the burning refuse bed at a temperature of 1100°C (1373 K), at elevation
12.5 m and at a velocity of 1.2 m/sec. At the end of the furnace, 25 m^3/
min of quench tank vapors leave the furnace at 300°C (573 K) and 0.1 m/sec.
The two gas flows leave through a vertical outlet flue at the top of the chamber   The entrance to the flue is at elevation 17.5 m and the flue cross-
sectional area is 65 m^2. Estimate the average velocity through the flue and
the possible peak velocity in consideration of buoyancy effects. Neglect the
flow area for the cold gases.

Mean velocity

$$\overline{v} = \frac{\text{Volumetric flow rate}}{\text{flue area}}$$

$$= \frac{6000}{65} \text{ m/min}$$

$$= 92.31 \text{ m/min} \quad \text{or } 1.54 \text{ m/sec}$$

Buoyancy-affected velocity from Eq. (146)

$$(u^2)_{hot} = (1.2)^2 + 2(9.807)(17.5 - 12.5)\left(\frac{1373}{573} - 1\right)$$

$$u_{hot} = 11.76 \text{ m/sec}$$

Clearly, the almost eight-fold increase in velocity will not fill the outlet
flue: there will be a "necking down" of the hot gas flow envelope in accord
with the basic continuity equation:

$$(\text{Density})(\text{velocity})(\text{area}) = \text{constant} = \text{mass flow rate} \tag{148}$$

Note also that for a "leaky" furnace, and an ambient temperature of
20°C (293 K):

$$u_{hot} = 19.05 \text{ m/sec}$$

$$u_{cold} = 9.68 \text{ m/sec}$$

Beyond these conclusions (which could allow the designer to estimate, e.g.,
erosion effects at the entrance to a boiler tube bank in the outlet flue) it
should also be evident that such accelerations will significantly shorten the
residence time of the burning/burned gases in the furnace chamber. Thus,
just as volumetric flow rate and flue area did not produce a good estimate of
exit velocity, neither will volumetric flow rate and chamber volume produce
a good estimate of mean residence time unless (1) the furnace is well sealed,
and (2) the system is of relatively uniform temperature.

Chapter 5

INCINERATION APPLICATIONS OF THE COMBUSTION PROCESS

This chapter goes beyond the basic framework of combustion theory developed in the preceding chapters to introduce practical waste incineration problems, components, and systems. This begins with the problem defining step of waste characterization. The discussion of incinerator systems reviews the basic components and systems reflecting modern practice in combustion-based waste management.

A. WASTE CHARACTERIZATION

The first step in solving waste management problems is to discard the belief that "waste" is so heterogeneous in its composition and so variable in its properties that problems cannot be defined, let alone solved. To be sure, waste streams exhibit great variability. The designer must, therefore, provide for more operating flexibility, reserve capacity and materials "stamina" than conventional process equipment. Nonetheless, development of an average waste composition and a range of probable excursions from that average provides the starting point for design.

Several properties of potential importance in waste characterization shown in Table 24 are drawn from Ref. 107. Although it is seldom necessary to characterize wastes in all the areas in Table 24, such a checklist can be useful in alerting the waste manager to processing constraints and opportunities or safety hazards.

In designing waste processing systems it will often be found that desired data are lacking. Many waste studies have demonstrated the large errors possible from desk-top estimates of the generation rate, composition,

TABLE 24

General, Physical, and Chemical Parameters of Possible Significance in the
Characterization of Solid Wastes[a]

General parameters	
Compositional weight fractions	Process weight fractions
Domestic, commercial, and institutional	Combustible
Paper (broken into subcategories)	Compostable
Food waste	Processable by landfill
Textiles	Salvageable
Glass and other ceramics	Having intrinsic value
Plastics	
Rubber	
Leather	
Metals	
Wood (limbs, sawdust)	
Bricks, stones, dirt, ashes	
Other municipal	
Dead animals	
Street sweepings	
Catch-basin cleanings	
Agricultural	
Field	
Processing	
Animal raising	
Industrial	

Mining/metallurgical

Special
Radioactive
Munitions, etc.
Pathogenic

Physical parameters

Total wastes	Solid wastes	Liquid wastes	Gaseous wastes
Size	Soluble (%)	Turbidity	Temperature
Shape	Suspendable (%)	Color	Pressure
Volume	Combustible (%)	Taste	Volume
Weight	Volatile (%)	Odor	Density
Density	Ash (%)	Temperature	Particulate (%)
stratification	soluble (%)	Viscosity data	Liquid (%)
Surface area	suspendable	specific gravity	
Compaction	Hardness	stratification	
Compactibility	Particle–size distribution	Total solids (%)	
Temperature	shape	soluble (%)	
Color	surface	suspended (%)	
Odor	porosity	settleable (%)	
Age	sorption	Dissolved oxygen	
Radioactivity	density	Vapor pressure	
Physical state	aggregation	Effect of shear rate	
total solids		Effect of temperature	
liquid		Gel formation	
gas			

TABLE 24 (Cont.)

General	Chemical parameters	
	Organic	Inorganic and elemental
pH	Soluble (%)	Moisture content
Alkalinity	Protein nitrogen	Carbon
Hardness (CaCO$_3$)	Phosphorus	Hydrogen
MBAS (methylene blue active substances)	Lipids	(P$_2$O$_5$ and phosphate)
BOD (biochemical oxygen demand)	Starches	Sulfur content
COD (chemical oxygen demand)	Sugars	Alkali metals
Rate of availability of nitrogen	Hemicelluloses	Alkaline-earth metals
Rate of availability of phosphorus	Lignins	Heavy metals
Crude fiber	Phenols	especially Mercury
Organic (%)	Benzene oil	Lead
Combustion parameters	ABS (alkyl benzene sulfonate)	Cadmium
Heat content	CCE (carbon chloroform extract)	Copper
Oxygen requirement	PCB (polychlorinated biphenyls)	Nickel
Flame temperature	PNH (polynuclear hydrocarbons)	Toxic materials
Combustion products (including ash)	Vitamins (e.g., B-12)	Chromium
Flash point	Insecticides (e.g., Heptachlor,	especially Arsenic
Ash fusion characterization	DDT, Dieldrin, etc.)	Selenium
Pyrolysis characterization	Precious metals	Beryllium
Toxicity		Asbestos
Corrosivity		Eutrophic materials
Explosivity		Nitrogen
Other safety factors		Potassium
Biological stability		Phosphorus
Attractiveness to vermin		

aSource: Ref. 107.

or properties of waste. It is strongly recommended, therefore, that the
results of especially commissioned waste surveys and analyses be incorporated into the problem definition phase of the design effort.

Careful consideration should also be given to the range of compositional
variation. For municipal waste, for example, seasonal changes in yard
waste and, paralleling local precipitation patterns, day to day fluctuations
in moisture content can be anticipated. In industry, seasonal shifts in production patterns or periodic housekeeping activity leads to variation.

Such changes in waste characteristics must be provided for in the design and operating protocols of waste processing systems. Even with such
relatively obvious foresight, however, the worst (live ammunition, cans of
flammable solvent, containers of toxic chemical, etc.) should be anticipated.
The cardinal rule in waste management design is to ask, "What happens
when... ?" rather than "What if... ?"

Although the analysis of the specific wastes to be processed is desirable, it is useful to have some general data for preliminary screening of
concepts. The data presented below meet this need. In general, these data
were generated by methodical sampling, segregation, weighing, and/or
analysis of the waste streams. More comprehensive data can be found elsewhere [108].

1. Solid Waste Composition

In this context, composition refers to the category of material (paper, glass,
etc.) in the waste streams. Compositional data are reported in this form
since the "analysis method," (visual categorization) is low-cost and can
rapidly and economically be applied to large quantities of waste. This latter
point is important if a meaningful characterization is to be made on a stream
which is grossly heterogeneous. Data in this form may be translated into
mean overall chemical compositions and the like by taking the weighted average of the chemical compositions of the specified components. Lastly,
data on a categorical basis are directly usable to estimate the potential for
materials recovery.

a. Mixed Municipal Refuse

In many instances, unfortunately, the waste stream of interest cannot
be directly sampled. Under such circumstances, data from other municipalities can be useful as an indicator of mean refuse composition. Generally,
municipal refuse is categorized as shown in Table 25.

An examination of refuse compositional data from across the United
States shows great variability, reflecting local practices regarding the

TABLE 25

Primary Constituents of Categories
of Mixed Municipal Refuse

Category	Description
Glass	Bottles (primarily)
Metal	Cans, wire, and foil
Paper	Various types, some with fillers
Plastics	Polyvinyl chloride, polyethylene, styrene, etc., as found in packing, housewares, furniture, toys and nonwoven synthetic fabrics
Leather, rubber	Shoes, tires, toys, etc.
Textiles	Cellulosic, protein, woven synthetics
Wood	Wooden packaging, furniture, logs, twigs
Food wastes	Garbage
Miscellaneous	Inorganic ash, stones, dust
Yard wastes	Grass, brush, shrub trimmings

wastes accepted at landfills or incinerators, seasonal effects (e.g., on yard waste quantities), economic level of the citizens, incorporation of commercial or industrial waste, etc. These data can be rationalized, however [109, 110], into an estimated national average composition.

These results are more useful if the moisture levels of the components are adjusted, category by category, to a moisture basis corresponding to the manufactured state of the materials entering the refuse storage bunker, that is, changing from the mixed or as-fired basis to as-discarded basis. The moisture contents shown in Table 26 were used to effect this basis shift.

Carrying out the moisture adjustment does not materially change the total moisture content of the refuse mix, only the distribution of moisture among the refuse categories. Discarded solid waste, as it is mixed together with other refuse materials, may either lose or absorb moisture. Food wastes, for example, may transfer significant quantities of moisture to paper and textiles. The as-discarded basis is useful in indicating the true relative magnitude of waste generation for the various categories, as the appropriate basis for estimating salvage potential, and as the basis for forecasting refuse generation rates and chemical and physical properties.

TABLE 26

Estimated Average Percent Moisture in Refuse on
an "As-discarded" and "As-fired" Basis[a]

Component	As-fired	As-discarded
Food wastes	63.6	70.0
Yard wastes	37.9	55.3
Miscellaneous	3.0	2.0
Glass	3.0	2.0
Metal	6.6	2.0
Paper	24.3	8.0
Plastics	13.8	2.0
Leather, rubber	13.8	2.0
Textiles	23.8	10.0
Wood	15.4	15.0

[a]From Ref. 109.

The seasonal and annual average compositions shown in Table 27 were derived from an analysis of over 30 data sets from municipalities throughout the United States. Using these data as defining a base year (1970), composition forecasts were prepared (Table 28) based on estimates of forecast consumption rates of paper, metal cans, and other consumer products. In the forecasts, consideration was given to the mean useful life of the products.

The data in Table 29 are given for comparison with the United States refuse composition estimates. For the United States refuse data, the decline in the use of coal for home heating is shown in the change in refuse ash content between 1939 and 1970.

b. Commercial and Industrial Waste

Solid wastes from commercial sources (stores, small buinesses) almost equal that generated in residences. Industrial waste generation can exceed the combination of residential and commercial wastes. Yet, with their importance apparently obvious, little published data exists.

TABLE 27

Estimated Average Municipal Refuse Composition,
1970 (Weight Percent, as Discarded)[a]

Category	Summer	Fall	Winter[b]	Spring	Annual average	
					As-discarded	As-fired
Paper	31.0	39.9	42.4	36.5	37.4	44.0
Yard wastes	27.1	6.2	0.4	14.4	13.9	9.4
Food wastes	17.7	22.7	24.1	20.8	20.0	17.1
Glass	7.5	9.6	10.2	8.8	9.8	8.8
Metal	7.0	9.1	9.7	8.2	8.4	8.6
Wood	2.6	3.4	3.6	3.1	3.1	3.0
Textiles	1.8	2.5	2.7	2.2	2.2	2.6
Leather, rubber	1.1	1.4	1.5	1.2	1.2	1.5
Plastics	1.1	1.2	1.4	1.1	1.4	1.4
Miscellaneous	3.1	4.0	4.2	3.7	3.4	3.6
Total	100.0	100.0	100.0	100.0	100.0	100.0

[a]From Ref. 110.
[b]For southern states, the refuse composition in winter is similar to that shown here for fall.

TABLE 28

Projected Average Generated Refuse Composition, 1970-2000[a]

Composition (wt %, as discarded)	1970	1975	1980	1990	2000
Paper	37.4	39.2	40.1	43.4	48.0
Yard wastes	13.9	13.3	12.9	12.3	11.9
Food wastes	20.0	17.8	16.1	14.0	12.1
Glass	9.0	9.9	10.2	9.5	8.1
Metal	8.4	8.6	8.9	8.6	7.1
Wood	3.1	2.7	2.4	2.0	1.6
Textiles	2.2	2.3	2.3	2.7	3.1
Leather, rubber	1.2	1.2	1.2	1.2	1.3
Plastics	1.4	2.1	3.0	3.9	4.7
Miscellaneous	3.4	3.0	2.7	2.4	2.1
Total	100.0	100.0	100.0	100.0	100.0

[a]From Ref. 110.

The information on industrial and commercial waste generation rates and compositions is difficult to obtain, correlate, and/or to generalize because of the following factors:

Manufacturing establishments, even those in the same type of business, may differ widely in their waste generating practices.

Most firms are reluctant to reveal production and related statistics for fear of the data being used to the competitive advantage of others.

Some firms are reluctant to provide information on waste volumes and composition for fear of it indicating noncompliance with pollution control regulations. Regardless of their wastes' pollution-related characteristics, there is also a tendency for firms to underestimate its quantity.

Some industrial activities are subject to seasonal fluctuations.

The extent of salvaging, recycling, sale to scrap dealers, or other reclamation of wastes differs greatly among manufacturers.

TABLE 29

A Summary of International Refuse Composition (Weight Percent, Mixed Refuse)[a]

	Ash	Paper	Organic matter	Metals	Glass	Miscellaneous
United States (1939)	43.0	21.9	17.0	6.8	5.5	5.8
United States (1970)[b]	0	44.0	26.5	8.6	8.8	12.1
Canada	5	70	10	5	5	5
United Kingdom	40-40	25-30	10-15	5-8	5-8	5-10
France[c]	24.3	29.6	24	4.2	3.9	14
West Germany[d]	30	18.7	21.2	5.1	9.8	15.2
Sweden	0	55	12	6	15	12
Spain[e]	22	21	45	3	4	5
Switzerland	20	40-50	15-25	5	5	—
Netherlands[f]	9.1	45.2	14	4.8	4.9	22
Norway (summer)	0	56.6	34.7	3.2	2.1	8.4

Norway (winter)	12.4	24.2	55.7	2.6	5.1	0
Israel	1.9	23.9	71.3	1.1	0.9	1.9
Belgium[g]	48	20.5	23	2.5	3	3
Czechoslovakia[h] (summer)	6	14	39	2	11	28
Czechoslovakia[h] (winter)	65	7	22	1	3	2
Finland	—	65	10	5	5	15
Poland	10–21	2.7–6.2	35.3–43.8	0.8–0.9	0.8–2.4	—
Japan (1963)	19.3	24.8	36.9	2.8	3.3	12.9

[a] From Refs. 111–113.
[b] From Table 27 (organic matter = yard and food waste; misc. = plastics, leather and rubber, wood, textiles, and miscellaneous).
[c] Paris (considered representative of national average).
[d] West Berlin.
[e] Madrid.
[f] The Hague.
[g] Brussels.
[h] Prague.

Many firms have themselves little understanding of and few records on their waste generation and characteristics.

Data for mixed commercial refuse and for a variety of industries have been prepared, however (Tables 30 and 31 and Ref. 108), but considerable care should be exercised in their application to any specific industrial establishment or geographic region.

2. Solid Waste Properties

The categorical composition is the starting point for the development of parameters of interest to the incinerator designer. Although the manipulation of gross categorical data to establish average chemical composition, heat content, and the like requires assumptions of questional accuracy, it is a necessary compromise. Typically, several (perhaps 1-3) tons of waste

TABLE 30

Composition of Commercial Refuse (Weight Percent, Mixed)

| Component | Commercial wastes | | Residential wastes[c] |
	Kentucky[a]	Michigan[b]	
Metal	10.6	6	8.6
Paper	60.4	57	44.0
Plastics	9.4	1	1.4
Leather, rubber			1.5
Textiles		1	2.6
Wood		2	3.0
Food waste	7.1	24	17.1
Yard waste		0	9.4
Glass	11.3	6	8.8
Miscellaneous	1.2	3	3.6
Total	100.0	100	100.0

[a] Source: Ref. 114 (sampling and analysis).
[b] Source: Ref. 115 (engineering estimates).
[c] Source: Table 27.

TABLE 31

Industrial Waste Composition[a]

SIC number	Component (wt %)									
	Paper	Wood	Leather	Rubber	Plastics	Metals	Glass	Textiles	Food	Miscellaneous
20 Food and kindred products										
Data points	30	30	30	30	30	30	30	30	30	30
Average	52.3	7.7	—	—	.9	8.2	4.9	0	16.7	9.2
Standard deviation	32.7	10.9	—	—	.4	3.7	2.8	0	29.9	21.1
Confidence limits	11.7	3.9	—	—	.1	1.3	1.0	0	10.7	7.5
22 Textile mills										
Data points	18	18	18	18	18	18	18	18	18	18
Average	45.5	—	0	—	4.7	—	—	26.8	—	—
Standard deviation	40.3	—	0	—	10.7	—	—	38.1	—	—
Confidence limits	18.6	—	—	—	4.9	—	—	17.6	—	—
23 Apparel products										
Data points	17	17	17	17	17	17	17	17	17	17
Average	55.9	—	0	0	—	0	0	36.5	1.35	—
Standard deviation	37.4	—	0	0	—	0	0	37.3	2.8	—
Confidence limits	17.8	—	—	—	—	—	—	17.7	1.3	—

TABLE 31 (Cont.)

SIC number	Component (wt %)									
	Paper	Wood	Leather	Rubber	Plastics	Metals	Glass	Textiles	Food	Miscellaneous
24 Wood products										
Data points	9	9	9	9	9	9	9	9	9	9
Average	16.7	71.6	0	0	0	—	—	0	0	7.8
Standard deviation	33.6	34.8	0	0	0	—	—	0	0	19.7
Confidence limits	22.0	22.7								12.9
25 Furniture										
Data points	7	7	7	7	7	7	7	7	7	7
Average	24.7	42.1	0	—	—	—	0	—	—	—
Standard deviation	12.3	16.2	0	—	—	—	0	—	—	—
Confidence limits	9.1	12.0								
26 Paper and allied products										
Data points	20	20	20	20	20	20	20	20	20	20
Average	56.3	11.3	0	0	—	9.4	—	—	—	14.0
Standard deviation	8.7	15.5	0	0	—	18.2	—	—	—	27.5
Confidence limits	3.8	6.8				8.0				12.1
27 Printing and publishing										
Data points	26	26	26	26	26	26	26	26	26	26
Average	84.9	5.5	—	0	—	—	0	—	—	—

Standard deviation	5.8	12.3	—	0	—	—	0	—	—	—
Confidence limits	2.2	4.7	—	0	—	—	0	—	—	—
28 Chemical and allied										
Data points	48	48	48	48	48	48	48	48	48	48
Average	55.0	4.5	—	—	9.3	7.2	2.2	—	—	19.7
Standard deviation	34.0	6.2	—	—	17.0	13.9	4.2	—	—	32.8
Confidence limits	9.6	1.7			4.8	3.9	1.2			9.3
29 Petroleum and allied										
Data points	5	5	5	5	5	5	5	5	5	5
Average	72.1	6.8	0	0	15.3	4.4	0	0	0	1.0
Standard deviation	35.7	4.4	0	0	30.7	5.2	0	0	0	1.3
Confidence limits	31.4	3.9			27.0	4.6				1.1
30 Rubber and plastics										
Data points	13	13	13	13	13	13	13	13	13	13
Average	56.3	5.2	0	9.2	13.5	—	0	—	—	—
Standard deviation	31.5	6.2	0	20.3	20.7	—	0	—	—	—
Confidence limits	17.2	3.4		11.0	11.3	—		—	—	—
31 Leather manufacturing										
Data points	3	3	3	3	3	3	3	3	3	3
Average	6.0	3.9	53.3	—	—	13.5	—	0	0	—
Standard deviation	4.2	5.4	47.3	—	—	19.2	—	0	0	—
Confidence limits	4.7	6.1	53.6	—	—	21.7	—	0	0	—

TABLE 31 (Cont.)

SIC number	Component (wt %)									
	Paper	Wood	Leather	Rubber	Plastics	Metals	Glass	Textiles	Food	Miscellaneous
32 Stone, clay, and glass										
Data points	16	16	16	16	16	16	16	16	16	16
Average	33.8	4.3	0	—	—	8.1	12.8	—	0	40.0
Standard deviation	37.5	8.4	0	—	—	24.8	29.6	—	0	44.8
Confidence limits	18.4	4.1		—	—	12.2	14.5	—		22.0
33 Primary metals										
Data points	12	12	12	12	12	12	12	12	12	12
Average	41.0	11.6	0	—	5.4	5.5	2.0	0	—	29.0
Standard deviation	27.4	12.4	0	—	9.8	7.8	4.3	0	—	40.0
Confidence limits	15.5	7.0		—	5.5	4.4	2.4		—	22.7
34 Fabricated metals										
Data points	36	36	36	36	36	36	36	36	36	36
Average	44.6	10.3	0	—	—	23.2	—	—	—	12.2
Standard deviation	37.7	20.8	0	—	—	34.5	—	—	—	31.0
Confidence limits	12.3	6.8		—	—	11.3	—	—	—	10.1
35 Nonelectrical machinery										
Data points	48	48	48	48	48	48	48	48	48	48
Average	43.1	11.4	—	—	2.5	23.7	—	0	—	—

Statistic										
Standard deviation	34.3	19.5	—	—	6.8	30.8	—	0	—	—
Confidence limits	9.7	5.5	—	—	1.9	8.7	—	—	—	—

36 Electrical machinery

Statistic										
Data points	19	19	19	19	19	19	19	19	19	19
Average	73.3	8.3	0	—	3.5	2.3	—	0	1.2	—
Standard deviation	24.4	10.1	0	—	7.0	3.5	—	0	2.4	—
Confidence limits	11.0	4.5	—	—	3.1	1.6	—	—	1.1	—

37 Transportation

Statistic										
Data points	8	8	8	8	8	8	8	8	8	8
Average	50.9	9.4	0	1.4	2.1	—	—	0	—	—
Standard deviation	34.2	6.3	0	1.5	2.9	—	—	0	—	—
Confidence limits	23.8	4.4	—	1.0	2.0	—	—	—	—	—

38 Scientific instruments

Statistic										
Data points	8	8	8	8	8	8	8	8	8	8
Average	44.8	2.3	0	0	6.0	8.4	—	0	—	—
Standard deviation	34.0	3.6	0	0	6.4	17.2	—	0	—	—
Confidence limits	23.6	2.5	—	—	4.4	11.9	—	—	—	—

39 Miscellaneous manufacturing

Statistic										
Data points	20	20	20	20	20	20	20	20	20	20
Average	54.6	13.0	—	—	11.9	5.0	—	—	—	8.1
Standard deviation	38.7	23.7	—	—	22.2	10.3	—	—	—	14.0
Confidence limits	17.0	10.4	—	—	9.7	4.5	—	—	—	6.1

[a]Source: Ref. 110; confidence limits are 95%.

from a 200 to 1000 ton/day waste flow are analyzed to produce a categorical composition. Then a still smaller sample is hammermilled, mixed, and a 500 mg sample is taken. Clearly, a calorific value determination on the latter sample is, at best, a rough reflection of the energy content of the original waste.

a. Chemical Analysis

Stepping from the categorical analysis to a mean chemical analysis provides the basis for stoichiometric calculations.

(1) Mixed Municipal Refuse

Chemical data for average municipal refuse components and for the mixed refuse shown in Table 27 are presented in Tables 32 and 33. These data were developed with the average refuse of 1970 in mind, but may be used to develop the composition of "future" refuse such as given in the projections in Table 28.

(2) Specific Waste Components

Data for specific waste components are given in Table 34. These data may be used when detailed categorical analyses are available or when one wishes to explore the impact of refuse compositional changes.

b. Bulk Density

In many smaller municipalities, weighing scales for refuse vehicles are not available. In such circumstances, data is often gathered on a volumetric basis. Reported bulk density data are presented in Table 35.

c. Thermal Parameters

Of great interest to the combustion engineer, once a weight-basis generation rate is established, is data on the heating value of waste components. In addition to those data given in Table 34, the information in Table 36 is presented.

Incineration is often considered for the disposal of wastewaters containing high concentrations of organic material. The contaminant concentration of these wastes is often determined as the chemical oxygen demand or COD. For this test, the oxygen uptake from a highly oxidizing chemical (often chromic acid) is used. The results are reported in milligrams (of oxygen) per liter. As noted in Table 37, the higher heating value for a wide variety of fuels is approximately the same per unit weight of oxygen: about

(Text continues on p. 220.)

TABLE 32

Estimated Ultimate Analysis of Refuse Categories[a]
(%, Dry Basis)

Category	C	H	O	N	Ash	S	Fe	Al	Cu	Zn	Pb	Sn	P**	Cl	Se	% Fixed carbon (dry basis)
Metal	4.5	0.6	4.3	0.05	90.5	0.01	77.3	20.1	2.0	—	0.01	0.6	0.03	—	—	0.5
Paper	45.4	6.1	42.1	0.3	6.0	0.12	—	—	—	—	—	—	—	—	Trace	11.3
Plastics	59.8	8.3	19.0	1.0	11.6	0.3	—	—	—	—	—	—	0.01	6.0	—	5.1
Leather, rubber	—	—	—	—	—	—	—	—	—	2.0	—	—	—	—	—	6.4
Textiles	46.2	6.4	41.8	2.2	3.2	0.2	—	—	—	—	—	—	0.03	—	—	3.9
Wood	48.3	6.0	42.4	0.3	2.9	0.11	—	—	—	—	—	—	0.05	—	—	14.1
Food wastes	41.7	5.8	27.6	2.8	21.9	0.25	—	—	—	—	—	—	0.24	—	—	5.3
Yard wastes	49.2	6.5	36.1	2.9	5.0	0.35	—	—	—	—	—	—	0.04	—	—	19.3
Glass	0.52	0.07	0.36	0.03	99.02	—	—	—	—	—	—	—	—	—	—	0.4
Miscellaneous	13.0*	2.0*	12.0*	3.0*	70.0	—	—	—	—	—	—	—	—	—	—	7.5

[a]From Ref. 21; (*) estimated (varies widely); (**) excludes phosphorus in $CaPO_4$.

TABLE 33

Ultimate Analysis of Annual Average 1970 Mixed Municipal Refuse[a]

Category	(wt %) "As-fired"	(wt %) "As-discarded"	(% moisture) "As-discarded"	Composition of average refuse (kg/100 kg dry solids)						
				Ash	C	H_2	O_2	S	V_2	Weight
Metal	8.7	8.2	2.0	10.13	0.50	0.067	0.481	0.0011	0.0056	11.19
Paper	44.2	35.6	8.0	2.74	20.70	2.781	19.193	0.0547	0.1368	45.59
Plastics	1.2	1.1	2.0	0.17	0.90	0.125	0.285	0.0045	0.0150	1.50
Leather, rubber	1.7	1.5	2.0	0.24	1.23	0.170	0.390	0.0062	0.0205	2.05
Textiles	2.3	1.9	10.0	0.08	1.10	0.152	0.995	0.0048	0.0523	2.38
Wood	2.5	2.5	15.0	0.09	1.43	0.178	1.260	0.0033	0.0089	2.96
Food waste	16.6	23.7	70.0	2.17	4.13	0.574	2.730	0.0248	0.2772	9.90
Yard waste	12.6	15.5	50.0	0.54	5.31	0.701	3.890	0.0378	0.3129	10.79
Glass	8.5	8.3	2.0	11.21	0.06	0.008	0.041	—	0.0034	11.32
Miscellaneous	1.7	1.7	2.0	1.62	0.30	0.046	0.278	—	0.0696	2.32
Total	100.0	100.0		28.99	35.66	4.802	29.543	0.1372	0.9022	100.00

Average Refuse Summary (As-fired Basis: 100 kg Average Refuse)

Component	wt %	Mol
Moisture (H_2O)	28.16	1.564
Carbon (C)	25.62	2.135
Hydrogen (H_2-bound)	2.65	1.326
Oxygen (O-bound)	21.21	1.326
Hydrogen (H_2)	0.80	0.399
Sulfur (S)	0.10	0.003
Nitrogen (N_2)	0.64	0.023
Ash	20.82	—
Total	100.00	

Higher heating value (water condensed): 2472 kcal/kg

Lower heating value (water as vapor): 2167 kcal/kg

[a] From Ref. 21.

TABLE 34

Proximate and Ultimate Analyses and Heating Value of Waste Components

Waste Component	Proximate Analysis (as-received) Weight %				Ultimate Analysis (dry) Weight %						Higher Heating Value (kcal/kg)		
	Moisture	Volatile Matter	Fixed Carbon	Non-Comb.	C	H	O	N	S	Non-Comb.	As Received	Dry	Moisture and Ash Free
Paper And Paper Products													
Paper, Mixed	10.24	75.94	8.44	5.38	43.41	5.82	44.32	0.25	0.20	6.00	3778	4207	4475
Newsprint	5.97	81.12	11.48	1.43	49.14	6.10	43.03	0.05	0.16	1.52	4430	4711	4778
Brown Paper	5.83	83.92	9.24	1.01	44.90	6.08	47.34	0.00	0.11	1.07	4031	4281	4333
Trade Magazine	4.11	66.39	7.03	22.47	32.91	4.95	38.55	0.07	0.09	23.43	2919	3044	3972
Corrugated Boxes	5.20	77.47	12.27	5.06	43.73	5.70	44.93	0.09	0.21	5.34	3913	4127	4361
Plastic-Coated Paper	4.71	84.20	8.45	2.64	45.30	6.17	45.50	0.18	0.08	2.77	4078	4279	4411
Waxed Milk Cartons	3.45	90.92	4.46	1.17	59.18	9.25	30.13	0.12	0.10	1.22	6293	6518	6606
Paper Food Cartons	6.11	75.59	11.80	6.50	44.74	6.10	41.92	0.15	0.16	6.93	4032	4294	4583
Junk Mail	4.56	73.32	9.03	13.09	37.87	5.41	42.74	0.17	0.09	13.72	3382	3543	4111
Food And Food Waste													
Vegetable Food Waste	78.29	17.10	3.55	1.06	49.06	6.62	37.55	1.68	0.20	4.89	997	4594	4833
Citrus Rinds and Seeds	78.70	16.55	4.01	0.74	47.96	5.68	41.67	1.11	0.12	3.46	948	4453	4611
Meat Scraps (cooked)	38.74	56.34	1.81	3.11	59.59	9.47	24.65	1.02	0.19	5.08	4235	6913	7283
Fried Fats	0.00	97.64	2.36	0.00	73.14	11.54	14.82	0.43	0.07	0.00	9148	9148	9148
Mixed Garbage I	72.00	20.26	3.26	4.48	44.99	6.43	28.76	3.30	0.52	16.00	1317	4713	5611
Mixed Garbage II	–	–	–	–	41.72	5.75	27.62	2.97	0.25	21.87	–	4026	5144
Trees, Wood, Brush, Plants													
Green Logs	50.00	42.25	7.25	0.50	50.12	6.40	42.26	0.14	0.08	1.00	1168	2336	2361
Rotten Timbers	26.80	55.01	16.13	2.06	52.30	5.5	39.0	0.2	1.2	2.8	2617	3538	3644
Demolition Softwood	7.70	77.62	13.93	0.75	51.0	6.1	41.8	0.1	<.1	0.8	4056	4398	4442
Waste Hardwood	12.00	75.05	12.41	0.54	49.4	6.1	43.7	0.1	<.1	0.6	3572	4056	4078
Furniture Wood	6.00	80.92	11.74	1.34	49.7	6.1	42.6	0.1	<.1	1.4	4083	4341	4411
Evergreen Shrubs	69.00	25.18	5.01	0.81	48.51	6.54	40.44	1.71	0.19	2.61	1504	4853	4978
Balsam Spruce	74.35	20.70	4.13	0.82	53.30	6.66	35.17	1.49	0.20	3.18	1359	5301	5472
Flowering Plants	53.94	35.64	8.08	2.34	46.65	6.61	40.18	1.21	0.26	5.09	2054	4459	4700

Lawn Grass I	75.24	18.64	4.50	1.62	46.18	5.96	36.43	4.46	0.42	6.55	1143	4618	4944
Lawn Grass II	65.00	-	-	2.37	43.33	6.04	41.68	2.15	0.05	6.75	1494	4274	4583
Ripe Leaves I	9.97	66.92	19.29	3.82	52.15	6.11	30.34	6.99	0.16	4.25	4436	4927	5150
Ripe Leaves II	50.00	-	-	4.10	50.46	5.95	45.10	0.20	0.05	8.20	1964	3927	4278
Wood and Bark	20.00	67.89	11.31	0.80		5.97	42.37	0.15	0.05	1.00	3833	4785	4833
Brush	40.00	-	-	5.00	42.52	5.90	41.20	2.00	0.05	8.33	2636	4389	4778
Mixed Greens	62.00	26.74	6.32	4.94	40.31	5.64	39.00	2.00	0.05	13.00	1494	3932	4519
Grass, Dirt, Leaves	21–62	-	-	-	36.20	4.75	26.61	2.10	0.26	30.08	-	3491	4994
Domestic Wastes													
Upholstery	6.9	75.96	14.52	2.62	47.1	6.1	43.6	0.3	.1	2.8	3867	4155	4272
Tires	1.02	64.92	27.51	6.55	79.1	6.8	5.9	0.1	1.5	6.6	7667	7726	8278
Leather	10.00	68.46	12.49	9.10	60.00	8.00	11.50	10.00	0.40	10.10	4422	4917	5472
Leather Shoe	7.46	57.12	14.26	21.16	42.01	5.32	22.83	5.98	1.00	22.86	4024	4348	5639
Shoe, Heel & Sole	1.15	67.03	2.08	29.74	53.22	7.09	7.76	0.50	1.34	30.09	6055	6126	8772
Rubber	1.20	83.98	4.94	9.88	77.65	10.35	-	-	2.00	10.00	6222	6294	7000
Mixed Plastics	2.0	-	-	10.00	60.00		22.60	-	-	10.00	7833	7982	8889
Plastic Film	3–20	-	-	-	67.21	9.72	15.82	0.46	0.07	6.72	-	7692	8261
Polyethylene	0.20	98.54	0.07	1.19	84.54	14.18	0.00	0.06	0.03	1.19	10,932	10,961	11,111
Polystyrene	0.20	98.67	0.68	0.45	87.10	8.45	3.96	0.06	0.02	0.45	9122	9139	9172
Polyurethane	0.20	87.12	8.30	4.38	63.27	6.26	17.65	0.21	0.02	4.38(a)	6224	6236	6517
Polyvinyl Chloride	0.20	86.89	10.85	2.06	45.14	5.61	1.56	5.99	0.14	2.06(b)	5419	5431	5556
Linoleum	2.10	64.50	6.60	26.80	48.06	5.34	18.70	0.08	0.40	27.40	4528	4617	6361
Rags	10.00	84.34	3.46	2.20	55.00	6.60	31.20	0.10	0.13	2.45	3833	4251	4358
Textiles	15–31	-	-	-	46.19	6.41	41.85	4.12	0.20	3.17	-	4464	4611
Oils, Paints	0	-	-	16.30	66.85	9.63	5.20	2.18	-	16.30	7444	7444	8889
Vacuum Cleaner Dirt	5.47	55.68	8.51	30.34	35.69	4.73	20.08	2.00	1.15	32.09	3548	3753	5533
Household Dirt	3.20	20.54	6.26	70.00	20.62	2.57	4.00	6.26	0.01	72.30	2039	2106	7583
Municipal Wastes													
Street Sweepings	20.00	54.00	6.00	20.00	34.70	4.76	35.20	0.14	0.20	25.00	2667	3333	4444
Mineral (c)	2–6	-	-	-	0.52	0.07	0.36	0.03	0.00	99.02	-	47	-
Metallic (c)	3–11	-	-	-	4.54	0.63	4.28	0.05	0.01	90.49	-	412	4333
Ashes	10.00	2.68	24.12	63.2	28.0	0.5	0.8	-	0.5	70.2	2089	2318	7778

Source: Refs. 38–40, 116–118.

aRemaining 2.42% is chlorine.

bRemaining 45.41% is chlorine.

cHeat and organic content from labels, coatings, and remains of contents of containers.

TABLE 35

Bulk Density of Mixed Wastes and Waste Components (kg/m^3)

	Unspecified average	Loose	Other	Range	Reference
Mixed residential wastes					
Garbage, kitchen waste	370	185		300–450	119
Mixed refuse with garbage		185	195[a]		119
Mixed refuse without garbage		140	280[a]		119
Residential	145			60–255	120
Residential in paper sacks	115			80–160	120
Wet residential	170			115–255	120
Damp residential	130			90–215	120
Dry residential	120			55–145	120
Bunkered refuse (8–11 m high)					
10% Moisture	155				121
20% Moisture	180				121
30% Moisture	225				121
40% Moisture	275				121
50% Moisture	345				121
Single-family dwelling	105				121

Multiple-family dwelling	110			122
Apartment house	150			122
Residential waste components				
Grass and trimmings		130		119
Metal cans		95		119
Unbroken glass and bottles		415		119
Broken glass		1190		123
Rags		115		124
Paper and cardboard		110	215[b]	119
Paper		140		124
Wet paper		165		124
Rubber		270		124
Bulky wastes				
Household bulky	100			125
Average bulky	190			125
Tree cuttings	135			125
Logs and stumps	400			119
Green logs		320	800[c]	121
Limbs and leaves		160	190[d]	119
Brush		30		119

TABLE 35 (Cont.)

	Unspecified average	Loose	Other	Range	Reference
Bulky wastes (Cont.)					
Furniture		50			119
Major appliances		180			119
Wood crates		110			119
Battery case and miscellaneous auto	715				119
Auto bodies		130			119
Wood pallets, driftwood	210				125
Tires and rubber products		240			119
Construction–demolition					
Mixed demo, nonburnable		1430			119
Mixed demo, burnable		360			119
Mixed const, burnable		255			125
Mixed construction I		965			119
Mixed construction II	160				125
Broken pavement, sidewalk		1520			119

Municipal		
Street dirt	1370	119
Alley cleanings	150	126
Street sweepings, litter	250	125
Catch-basin cleaning	1445	119
Sewage-sludge solids	1040	119
Sewage screenings	950	119
Sewage skim (grease)	950	119
Industrial waste		
Sawdust	290	123
Bark slabs	400	123
Wood trimmings	580	123
Wood shavings	240	123
Mixed metals	120	123
Heavy metal scrap	2410	119
Light metal scrap	800	119
Wire	320	119
Dirt, sand, gravel	1445	119

TABLE 35 (Cont.)

	Unspecified average	Loose	Other	Range	Reference
Industrial waste (Cont.)					
Cinders		900			124
Fly ash		1285			119
Cement industry waste		1425			119
Other fine particles		965			119
Oils, tars, asphalts		965			119
Mixed sludges (wet)		1190			123
Chemical waste, dry		640			119
Chemical waste, wet		965			119
Leather scraps		180			123
Shells, offal, paunch, fleshings		300			123
Textile wastes		180			123

Rubber	300	[123]
Plastics	30	[123]
Agricultural waste		
Pen sweepings	650	[119]
Paunch manure	1030	[119]
Other meat-packing waste	1030	[119]
Dead animals	355	[119]
Mixed vegetable waste	355	[123]
Mixed fruit waste	355	[123]
Beans or grain waste	775	[119]
Potato-processing waste	670	[119]
Chaff	60	[123]
Mixed agricultural	565	[119]

[a] Dumped packer truck density, expanded from 215 kg/m^3 in packer truck.
[b] Compacted in packer truck.
[c] Density of wood.
[d] Chipped.

TABLE 36

Higher Heating Value of Refuse Components
(kcal/kg)[a]

	Ash content (dry basis)	Higher heating value (kcal/kg)	
		Dry basis	Dry, ash–free
Paper, paper products			
Books, magazines	24.05		4198
Cardboard			4652
Mixed paper I	6.55		4488
Mixed paper II	4.02		4283
Waxed paper			6111
Waxed cartons			6667
Food, food wastes			
Garbage (California)	6.53		4044
Cooking fats			9000
Sugar (sucrose)	0.00	3943	3943
Starch	0.00	4179	4179
Coffee grounds		5588	
Corn on the cob		4500	
Brown peanut skins		5795	
Oats		4443	
Wheat		4184	
Castor oil	0.00	8861	8861
Cottonseed oil	0.00	9400	9400
Trees, wood, brush, plants			
Brush (California)	8.87		4732
Excelsior	0.77		4792
Greens (California)	10.57		4396

TABLE 36 (Cont.)

	Ash content (dry basis)	Higher heating value (kcal/kg)	
		Dry basis	Dry, ash–free
Trees, wood, brush, plants (Cont.)			
Lignin			5850
Wood (California)	1.13		4678
Pitch		8406	
Wood (Washington, D.C.)	1.65		4925
Sawdust (pine)		5376	
Sawdust (fur)		4583	
Wood, beech (13% H_2O)		3636	
Wood, birch (11.8% H_2O)		3717	
Wood, oak (13.3% H_2O)		3461	
Wood, pine (12.2% H_2O)		3889	
Domestic wastes			
Linoleum	27.4		6361
Rags, linen		3962	
Rags, cotton			4000
Rags, silk		4662	
Rags, wool			5444
Rags, mixed	2.19		4189
Rags, cellulose acetate	2.19		4444
Rags, nylon			7328
Rags, rayon			416
Rubber	48.0		7111
Shellac			7544
Asphalt			9533

TABLE 36 (Cont.)

	Ash content (dry basis)	Higher heating value (kcal/kg)	
		Dry basis	Dry, ash-free
Plastics			
Phenol formaldehyde			6217
Polyethylene			11083
Polypropylene			11083
Polystyrene			9906
Polyurethane foam			5700
Styrene-butadiene copolymer			9833
Vinyl chloride/acetate copolymer			4906
Metals			
Aluminum (to Al_2O_3)			7417
Copper (to CuO)			603
Iron (to $FeO_{.947}$)			1200
Iron (to Fe_3O_4)			1594
Iron (to Fe_2O_3)			1756
Lead (to PbO)			250
Magnesium (to MgO)			5911
Tin (to SnO_2)			1169
Zinc (to NzO)			1275

[a] Source: Refs. 40, 127, 128; see also Table 34.

1360 kg of air per million kilocalories heat release. This equation can be used to estimate the heat content of such aqueous wastes. Specifically,

$$[\text{COD (mg/liter)}] \times [3.5 \times 10^{-3}] = \frac{\text{kcal}}{\text{liter}} \text{ (approx.)} \tag{149}$$

TABLE 37

Theoretical Air Requirements of Municipal Refuse and Other Fuels

Fuel	kg Air/kg fuel	kg Air/MM kcal[a]	sm^3 Air/M kcal
Refuse	3.22	1303	1.10
Wood	3.29	1265	1.07
Peat	2.33	1300	1.10
Lignite	5.27	1343	1.13
Sub-bituminous B	7.58	1332	1.12
Bituminous high volatile	9.08	1336	1.13
Bituminous volatile	10.99	1368	1.15
Anthracite	9.23	1496	1.26
Fuel oil	13.69	1350	1.14
Methane	17.26	1300	1.10

[a]MM = millions, M = thousands.

It should be noted that some classes of organic compounds (e.g., many aromatic compounds) are incompletely oxidized in the COD test procedure. Thus, estimates of combustion parameters from COD determinations may be in error (on the low side) depending upon the concentration of these chemically refractory materials.

Another estimation of waste heating value may be obtained from the ultimate analysis using Eq. (8).

Lastly, it should be noted that the approximate equivalence of heating value with air requirement can be used to estimate heating value from combustion stoichiometry or air requirement from bomb calorimeter data on heating value.

Example 17. (1) Estimate the theoretical air requirement and the heating value for methane (CH_4). (2) Also, estimate the air requirement for a waste of unknown composition which, from bomb calorimeter tests, has a heating value of 2900 kcal/kg.

1. Methane will burn according to the stoichiometric relationship

$$CH_4 + 2O_2 \longrightarrow CO_2 + 2H_2O$$

$$\frac{\text{kg Air}}{\text{kg Fuel}} = \frac{2 \text{ mol O}_2}{\text{mol CH}_4} \times \frac{\text{mol CH}_4}{16 \text{ kg CH}_4} \times \frac{100 \text{ mol air}}{21 \text{ mol O}_2} \times \frac{29 \text{ kg air}}{\text{mol air}}$$

$$= 17.3$$

$$\text{Approximate heating value} = \frac{17.3 \text{ kg air}}{\text{kg fuel}} \times \frac{10^6 \text{ kcal}}{1360 \text{ kg air}}$$

$$= 12,720 \text{ kcal/kg}$$

(actual: 13,272 kcal/kg, a 4% error).

2. Based on the approximate equivalence of 1360 kg of air with one million kcal, the estimated stoichiometric air requirement of the waste is

$$\frac{\text{kg Air}}{\text{kg Fuel}} = \frac{2900 \text{ kcal}}{\text{kg fuel}} \times \frac{1360 \text{ kg air}}{10^6 \text{ kcal}} = 3.94$$

d. Municipal Refuse as a Fuel

Electric power plants and industrial combustion systems represent a rich source of data and proven design experience for understanding and improving incineration systems. To use these data, however, it is necessary to appreciate the similarities and the differences between municipal refuse and its associated combustion parameters and those for other fuels.

(1) Heat Content

Municipal refuse, though quite different from high-rank coals and oil, has considerable similarity to wood, peat, and lignite (Table 38). It would be reasonable, therefore, to seek incinerator design concepts in the technology developed for the combustion of the latter materials. Refuse does, however, have distinguishing characteristics, such as its high total ash content, which may require more extensive ash handling equipment. Excluding the massive ash (metal and glass), however, the fine ash content (that capable of being suspended in the flue gas) is only 5.44%, suggesting that incinerators may require less efficient particulate air pollution control devices for comparable combustion situations. Clearly, refuse is also a lower energy fuel than most conventional solid fuel alternatives.

(2) Density

Refuse density is similar to that of wood, peat, and lignite (Table 39). Since, in grate fired systems, coal and refuse furnace retention times are comparable (40-50 min), refuse bed depths would be much greater than those for coal at comparable grate burning rates. A typical coal stoker, for

TABLE 38

Heat Content of Municipal Refuse and Other Fuels

Fuel	Moisture (%)	Ash (%)	Volatile (Dry basis) (%)	(Higher Heating Value) (kcal/kg)				As-fired (Lower Heating Value) (kcal/kg)
				As-fired	Ash-free	Dry basis	Dry, ash-free basis	
Refuse	28.16	20.82	62.3	2470	3120	3440	4845	2185
Wood	46.9	1.5	78.1	2605	2645	4910	5050	2170
Peat	64.3	10.0	67.3	1800	2000	5030	6995	1330
Lignite	36.0	12.1	49.8	3925	4465	6130	7560	3585
Sub-bituminous B	15.3	6.7	39.7	5690	6100	6720	7200	5375
Bituminous-high volatile B	8.6	8.4	35.4	6800	7420	7440	8190	6520
Bituminous-volatile	3.6	4.9	16.0	8030	8450	8330	8780	7775
Anthracite	4.5	14.4	7.4	6170	7205	6460	7600	6010
Methane	0	0	100.0	13,275	13,275	13,275	13,275	11,975
#2 Fuel oil	0	0	100.0	10,870	10,870	10,870	10,870	10,210
#6 Fuel oil	1.5	0.08	100.0	10,145	10,300	10,300	10,375	9,600

TABLE 39

Density of Municipal Refuse and Other Fuels

Fuel	Density (as–fired kg/m^3)	kcal/m^3	Density (ultimate) kg/m^3)	Mean void fraction (%)
Refuse	273 (av)	676	963 (est)	72
Wood (chips)	280	730	562 (av)	50
Peat	400 (est)	712	802	50
Bituminous coal (sized)	802	5783	1405	43
#6 Fuel	987	10,010	987	0
Methane	0.67	9	N/A	N/A

example, burns at the rate of 9300 to 12000 kcal hr^{-1} m^{-2}. This corresponds, for the bituminous coal energy density and for typical stoker speeds, to a bed height of 12.7 to 17.7 cm. In order to provide comparable specific burning rates for refuse, bed heights from 110 to 150 cm are required.

Table 38 indicates the average density of the municipal refuse in comparison with other fuels. The "ultimate density" for refuse was calculated as the weighted sum of the densities of the refuse components. Comparing the ultimate density with the as–fired density yields an approximate void fraction which is much higher for refuse than for any other fuel shown. The high void fraction, aside from reaffirming the need for a deep bed, also indicates the difficulties in obtaining uniform refuse distribution to avoid "blow-holes" or open spaces on the grate.

(3) Air Requirements

Although the stoichiometric air requirement per million kilocalories for refuse is in line with that for other fuels, refuse has a relatively low air requirement per kilogram of fuel (Table 39). Since refuse is now burned at bed depths which approximate coal heat release rates (kcal hr^{-1} m^{-2}), the air rates (m^3 sec^{-1} m^{-2}) could, in principle, be similar to those used for coal stokers. In practice, however, much higher air rates are used in incinerators to effect drying and to temper ultimate combustion temperatures.

In summary, municipal refuse burning practices should be comparable with those for low rank carbonaceous solid fuels such as wood, peat, and lignite. Although incinerator designs reflect a portion of the technology developed to burn these other materials, application of spreader stoker and suspension burning techniques and some aspects of furnace design and fuel feeding have not yet found their way into conventional incinerator practice. However, refuse does present some unique problems to the furnace designer, particularly as a result of its relatively low energy density, high ash content, and high moisture content.

B. INCINERATION SYSTEMS

The following discussions are intended to describe contemporary incineration practice for municipal and industrial incineration systems. In a few instances, particularly flue gas temperature control in refractory systems, some additional process analysis methodology is presented.

The predominate concern here is on municipal incinerators. Because of the wide variation in waste character in industrial situations, most systems require designs where materials handling, construction materials selection, effluent gas quality goals, and other aspects are highly tailored to the problem at hand. The complexity of the resultant designs and the many new dimensions in combustor configuration (e.g., fluidized beds) are generally beyond the scope of this volume.

1. The Municipal Incinerator

The objective of this section is to answer the difficult question: What is a municipal incinerator? Unlike the chemical or manufacturing industries, the incineration business employs a wide variety of designs to do the same job. Rarely are two United States plants built to precisely the same design. This individuality reflects both the growth of incineration technology in recent years and the large number of basic design parameters which are so poorly understood that they bend readily to the personal prejudices of the design engineer.

Thus an incinerator can be many things. The paragraphs which follow [21] outline very briefly the options in incinerator design. Any one of the pages or even paragraphs would require expansion to a chapter or book in its own right if the subject were dealt with at a level of detail fully supporting a decision making process. It is hoped, however, that after reading this section the reader will have familiarity with many of the terms involved in describing an incinerator and with the wide variability which is possible between systems.

a. Collection and Delivery of Refuse

In the incinerator system, the handling of refuse begins with the delivery of materials to the site. While in the United States most refuse is delivered to the site in motor vehicles, other countries sometimes use railroads or water transportation. Delivery vehicles include open dump trucks, commercial vehicles, private cars, and, most importantly, trucks with equipment for compression and densification. In some instances, 50 to 70-m^3 compaction truck trailers are employed to ferry refuse from centrally located transfer stations to a distant incinerator site. Such transfer stations are serviced by the smaller collection vehicles described above.

The refuse loaded into the original collection vehicles usually has a density of about 80 to 240 kg/m^3, depending on the nature of the refuse and how it has been packed at the refuse generation point. Refuse can be loosely placed in collection receptacles, manually stamp-packed, or, more recently, compacted at the site by compression and/or baling devices, sometimes after being ground or shredded. A power compaction unit can compress the refuse at the generation site (such as at commercial establishments, hospitals, apartments, hotels, etc.) to a density of about 500 to 1000 kg/m^3. However, most refuse loaded into vehicles at the collection site is compacted in the truck body to 250 to 500 kg/m^3 by means of mechanical and/or hydraulic systems employing vehicle motor power. The latter compaction is employed in most municipal and private collection vehicles to increase the payload capacity of the vehicle.

b. Refuse Handling and Storage

Refuse is most often received and stored in a pit below ground level. A traveling crane with a bucket is used to pile the refuse for storage and to move it away from the unloading area so that the pit can accommodate additional refuse. At the same time, the crane and bucket can be used to feed the incinerator furnace. Generally, the pit is sized to hold the quantity of refuse that can be burned in 2 to 3 days.

Three types of buckets are in common use for handling municipal refuse: the clamshell bucket, the grapple bucket, and the orange peel bucket. The clamshell and grapple buckets can be equipped with teeth to assist in digging into the refuse for a full bucket. The orange peel bucket does not require teeth to grab the refuse, even for unusual or large pieces. However, with the orange peel, it is difficult to clean the floor of the receiving pit, while the clamshell and the grapple are well suited to cleaning the pit, providing the teeth are removed. A steel plate is sometimes used to cover the teeth when cleaning the floor of the pit. Pit cleaning is important for sanitation, good housekeeping, and elimination of odors from decomposing refuse.

In some cases, a paved floor is used instead of a receiving pit; the refuse is deposited on the floor by the collection vehicles. The refuse can then be pushed onto a wide conveyor for loading either into the feed hopper of the incinerator furnace or to shredding or compacting devices. Alternatively, a front-end loader or bulldozer blade can be used to push the refuse through an opening in the floor to feed the incinerator furnace.

While open receiving areas are possible, they are infrequently used. The receiving area should preferably be closed to protect the refuse from blowing in the wind and causing a nuisance, to control dust and odors, and to prevent snow and ice from interfering with the movement of vehicles. Closure of the refuse receiving area also assists in refuse moisture control by keeping rain off and also preventing excessive drying. Moisture stability is helpful in obtaining uniform combustion temperatures in the furnaces.

Most incinerators do not have a scale to weigh the incoming refuse. However, for plant performance evaluation and for toll disposal, it may be necessary to know the weight of the refuse. Various types of scales are available. Some are manually operated with the weights recorded by the scale operator. Others can furnish a printout of the gross, tare, and net weights, together with the license or code number of the vehicle as supplied by the weighmaster. Newer installations may use a computerized scale for automatic weighing: the driver of the incoming vehicle inserts a coded card to supply the vehicle code number and the tare weight of the vehicle, and the automatic system prints the gross weight and net weight, as well as the date and time of the delivery.

c. Size Control and Salvage

In the past, oversize or otherwise undesirable refuse has been controlled by exclusion at the tipping location or by selective removal from the pit by the crane operator. Under supervision, bulky articles can be deposited in a specific area of the pit, or they can be carried directly to the residue disposal site without passing through the incinerator furnace. The latter method is often used for bulky metal items such as refrigerators, stoves, and large metal parts, which contain little or no combustible material. If the item contains biodegradable material it is cleaned before it is taken to the residue disposal site. Combustible oversize items are preferably broken up and incinerated.

Most refuse can be reduced in size by shearing or shredding. Shearing is a controlled method for reducing all refuse to a given maximum size by passing it through a shear or series of shears, or through a single or multiple guillotine. Shredding, which is rapidly coming into increased use, commonly makes use of horizontal or vertical shaft hammermills. Coarse shredding (say to a top-size of 25 cm) permits the removal of ferrous metal with magnetic drums or belts. Finer shredding, although necessary for applications where the refuse will be burned in suspension, presents problems when burning on conventional incinerator grates.

d. Incinerator Feed Systems

The feeding of the refuse may be either batch or continuous, although the recent trend has been toward the use of continuous firing to improve process control. Batch feeding of refuse directly into the furnace is done, in most cases, with a clamshell bucket or grapple attached to a traveling crane; the rate of feed is controlled by the time cycle and the degree of bucket loading. In a few plants, a front-end loader operating on a paved floor charges the furnace.

A ram-type feeding device is sometimes used for controlled feeding. With such a system, either the ram can clear the hopper at each stroke or an oversized hopper can be filled with refuse and the ram used to shear a horizontal section of refuse at selected intervals. The ram feeder provides an air seal at the feed to the furnace, an improvement over the bucket or the front-end loader systems of batch feeding, which usually let in undesirable quantities of cold air, as well as releasing occasional puffs of flames or hot gases while the charging gate is open. The inrush of cold air can be detrimental to the inside refractory walls of the furnace and can cause smoke evolution by cooling and quenching the burning process. While it is possible to use a two-gate airlock during batch feeding to keep the air our, such devices are seldom used because of the increased height and cost and extra maintenance involved in the additional mechanical equipment. However, from a technical standpoint, an air seal should be used.

Newer designs for incinerator systems nearly always specify continuous feeding of refuse to the incineration furnace (Fig. 37). Continuous feeding can be accomplished by means of a hopper and a gravity chute, a mechanical feeder, such as a pusher, ram, rotary feeder, or the like, which can be filled directly from a hopper supplied with refuse by a bucket and crane; a front-end loader from a feeding floor; a conveyor transporting the refuse from the receiving area; or an air injection system (shredding with suspension burning). The most frequently used system is the hopper and gravity chute. A rectangular hopper receives the refuse delivered by the crane and bucket. The bottom of the hopper terminates in a rectangular section chute leading downward to the furnace grate or other feeder conveyor at the entrance to the furnace chamber itself. This gravity chute is either of uniform cross-section or flared slightly toward the lower end to minimize bridging. It is either refractory lined, made of water cooled steel plate, or both, to prevent heat damage. Fires in the chute may be caused by faulty chute design (too short or too narrow), by the presence of large or irregular refuse which can provide flues for combustion gases and flame to ascend the chute, or by having insufficient material in the chute.

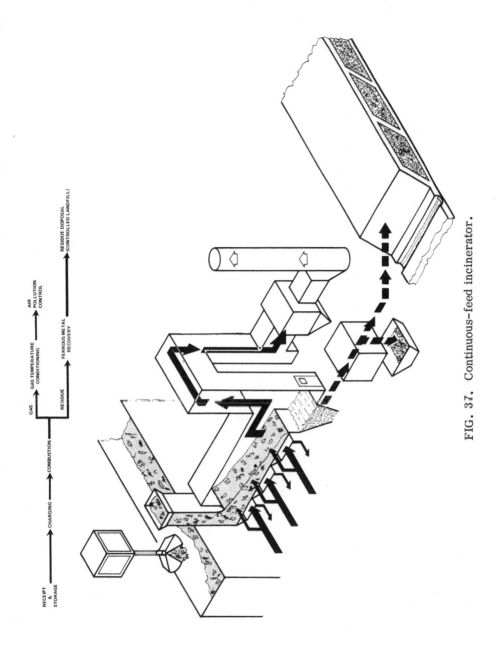

FIG. 37. Continuous-feed incinerator.

If suspension burning is to be employed in the incinerator furnace, the
refuse should be prepared by suitable shredding or grinding, and the most
desirable feeding method is air injection. Suspension burning has been used
successfully in waterwall boilers for the generation of steam from waste
products such as wood bark, bagasse, and similar materials. In con-
formance with practices in the fossil fuel fired boiler industry, air injection
of shredded refuse can be used for corner firing, for spreader firing, or
for "cyclone" firing, all of which are in commercial use for the generation
of steam from powdered or crushed bituminous coal. With suspension
burning, a burnout grate is provided at the bottom of the waterwall furnace
chamber to permit completion of combustion of larger particles or slow
burning materials.

e. Grates and Hearths

Nearly all incineration furnaces employ either a refractory hearth to
support the burning refuse or a variety of grate types which stoke or mix the
refuse during the combustion process in various ways depending upon the
type of grate or stoker. Suspension burning is the only process that does
not necessarily require a hearth or grate, since most of the refuse is oxidized while in suspension in furnace gases, but even for suspension burning
a burnout grate is often provided at the bottom of the furnace chamber to
assure complete burnout of combustibles in the residue. There are many
different types of hearth or grate, each of which has its own special features.

(1) Stationary Hearth

Those incinerator furnace systems which operate without grates include
the stationary hearth, rotary hearth, and rotary kiln. The stationary hearth
is usually a refractory floor to the furnace, which may have openings for
the admission of air under slight pressure below the burning material on the
hearth, in the manner of the blacksmith's forge. In the absence of underfire
air ports, air is admitted along the sides or from the top of the furnace, and
combustion proceeds in the same manner as in a bonfire, but with improved
conditions due to the reradiation of heat from the surrounding furnace walls
and roof. Unless the refuse being processed provides a porous burning mass,
as may be the case with bulky refuse, it is necessary to provide manual
stoking with slice bars to stir the mass of refuse in order to achieve a reasonable degree of burnout.

Stationary hearth furnaces are used for most commercial and smaller
industrial incinerators. They are also used in crematories and for hospital
wastes, when assisted with auxiliary gas or oil burners to maintain the furnace temperature above 650 to 900° C in the presence of adequate quantities
of oxygen from air, well-dispersed throughout the gas, to ensure complete
oxidation of combustible solids and vapors for the elimination of smoke and

odors. This may require auxiliary fuel burners in a secondary combustion chamber.

Increasing numbers of bulky refuse incinerators [117,118,121] have a refractory chamber with a refractory floor equipped with a number of air inlet gratings, similar to floor drains. One side or end of the chamber is connected to the hot combustion chamber of another incinerator installation. If not, a secondary combustion chamber must be installed prior to the breeching and stack to assure burnout of combustible gases and particulate. A refractory door permits the opposite side of the combustion chamber to be opened for access. A special extended bulldozer blade can be used to load the hearth with refuse from a paved platform outside the chamber and to drag out ash residue. Alternatively, it is possible to remove the ash residue by raking with a hoe or by pushing the ash residue beyond the hearth into a depressed area from which the ash can be removed.

(2) Rotary Kiln

The rotary kiln type of hearth has been used for several hundred years in the pyro-processing industry to move solids in and out of high temperature combustion zones and to mix them during combustion by rotating them. The steel cylindrical shell is lined with abrasion-resistant refractory to prevent overheating of the metal, unless special provisions are made for air cooling or water cooling. The kiln is inclined toward the discharge end, and the movement of the solids being processed is controlled by the speed of rotation.

There has been little use of the rotary kiln for municipal incinerator furnaces except after the burning of refuse on a multiple-grate system. Since the rotary kiln normally requires all of the air for combustion to enter the unit with the refuse at the feed end, the flow of combustion air is a deterrent to the temperature increase necessary for rapid ignition. The few rotary kilns that have been used for incineration of refuse have been relatively short, perhaps four to five times the diameter and are commonly provided with a bypass duct to reduce the gas flow (and particulate entrainment) through the kiln.

(3) Stationary Grates

Stationary grates have been used in incinerator furnaces for a longer time than any grate system except the stationary hearth, which was used in incinerators and crematories prior to the Middle Ages. The original stationary grates were probably metal bars or rails supported in the masonry side walls of the furnace chamber. Subsequently, these bars were replaced with cast metal or fabricated metal grates with provision for rotating the grate sections to permit dumping of the ash residue. While some stoking action can be obtained by shaking the grate, as in partial dumping, stationary grates normally require manual stoking with a slice bar to stir the burning bed of refuse in order to obtain a reasonably complete burnout of

ash residue. Such stationary grates are still in use in many of the older
cell-type incinerators. The latter have usually been constructed with a
multiplicity of furnace sections or cells, with an opening above each cell
for charging.

(4) Mechanical Grates: Batch Operations

Mechanically operated grates installed in batch-type furnaces were a
natural evolution from the stationary grate furnaces. Although batch-type
incineration furnaces have given way to continuous furnaces for new, large
installations, many of the new small-capacity incinerators still utilize
batch-fed furnaces, either with stationary or intermittently operated grates
or without grates, the latter in small commercial and industrial installations.

(a) Cylindrical Furnace Grates. In the circular batch furnace the grates
form annuli inside the vertical cylindrical walls of the furnace. A solid
grate or "dead plate" covers the central area of the annulus. A hollow rotating hub with extended rabble arms rotates slowly above the circular dead
plate to provide mechanical stoking or mixing. A center post provides the
bearing support of the hub and rabble arms. The rotating hub is covered
with a hemispherical "cone" with one or more consecutively smaller cones
stacked on top of the first one. Forced air (called "cone air") for combustion is supplied through the hub to the hollow rabble arms and thence through
openings in the arms to the space just above the dead plate. Additional cone
air is supplied to each of the cones, to cool the metal. The annular grate
area is divided into pairs of keystone shaped segments; each pair is arranged
to open downward for dumping the ash residue into the ash hopper below.
These segmental grates are either hand operated or hydraulically operated
under manual control. It is frequently necessary to poke and slice the ash
residue using the access doors around the furnace, in order to clean the
fires and to assist in the dumping of the ash residue.

When circular grates are used, the cylindrical furnace is fed intermittently or batchwise through an opening in the top at the centerline of the
furnace. A charging door is used to close the opening between additions of
refuse. In the usual method of operation, refuse is placed in the hopper
above the charging door or gate preparatory to the next charging cycle. At
a signal from the operator, the charging gate is opened, the refuse drops
into the bonfire of previously ignited refuse, and the charging gate closes
immediately after the hopper is emptied. Additional refuse is charged in
repeated cycles at the discretion of the operator until the accumulated ash
residue in the furnace requires dumping through the circular grates.

(b) Rectangular Batch Furnace Grates. Mechanically operated grates in
rectangular batch operated incinerator furnaces include reciprocating (pusher)
grates, and rocking grates. The grates are installed in a slightly inclined

position from the horizontal, with the lower end of the incline at the ash discharge point. With these grates, the furnace is fed intermittently through an opening in the top and at the higher end of the grate, and the fresh refuse is deposited over the bonfire of previously ignited refuse.

As the burning continues, the grates are operated under manual control to move the burning bed of refuse toward the discharge; with manual control to prevent (ideally) the discharge of residue that has not been completely burned. In some instances a dump grate is installed at the ash discharge point to hold back ash residue that is still burning, with manual operation of the dump grate after the accumulated ash has been completely burned. At times it is necessary to manually stoke or slice the fire in order to spread the burning refuse over the grate or remove larger pieces of metal or clinkers that interfere with the desired flow of burning refuse across the grate.

(5) Mechanical Grates: Continuous Operations

Mechanical constant-flow grates have been and are being used in most of the newer continuous burning incinerators. This constant-flow grate feeds the refuse continuously from the refuse feed chute to the incinerator furnace, provides movement of the refuse bed and ash residue toward the discharge end of the grate, and does some stoking and mixing of the burning material on the grates. Underfire air passes upward through the grate to provide oxygen for the combustion processes, while at the same time cooling the metal portions of the grate to protect them from oxidation and heat damage. Typical grate designs correspond to an average heat release rate of 13,500 kcal m^{-2} min^{-1}. Clearly, the actual rate in different portions of the grate differs widely from this average.

(a) <u>Reciprocating Grate</u>. The reciprocating or pusher grate is installed stepwise in rows on a slight downward incline toward the discharge; the rows move alternately to convey the refuse from the feed chute through the combustion area to the ash hopper. Although there is some stoking action as a result of the separate motion of the alternate rows of grates, additional stoking and mixing (breaking open packed refuse masses) are obtained by providing a drop-off for tumbling from one grate section to the next. Two, three, or four grate sections are commonly included in this type of grate for a continuous flow incinerator.

The reverse acting reciproacting grate is also inclined downward, through at a steeper angle, toward the discharge end. In this system, the cast metal grate elements reciprocate uphill against the downward flow of refuse, thereby producing some rolling of the burning material as a result of the upward reverse thrust. Again, these grates form a steplike configuration rather than a steady incline, so that additional mixing is obtained as the refuse tumbles from one level to the next.

(b) Rocking Grate. The rocking grate also slopes downward from the feed towards the discharge end, with two, three, four, or more grate sections installed in series with little or no (e.g., <3 cm) drop-off between grate sections. The rocking grate includes a multiplicity of grate sections or segments which are approximately quarter-cylindrical, and which include 20-30% open area to pass siftings to a hopper or conveyor system mounted below the grate and to admit undergrate air. Alternate rows of grate sections are rotated approximately 90° about the edge toward the discharge of the grates, with the grate face rising up into the burning mass and thus breaking it up and thrusting it forward toward the discharge. These grate sections each rotate back to a rest position and alternate sections rotate as before, causing a similar stoking action and pushing of the refuse bed forward. This two-cycle action is similar to the previously described pusher and reverse acting grates, except that it seems to mix the material more effectively. Further, it redistributes the refuse and ash on the grate to cover local areas where burning has left some grate surface exposed.

(c) Vibrating, Oscillating, and Impact Grates. Another incinerator grate classification consists of the vibrating or oscillating grate and the impact grate. The vibrating or oscillating grate is mechanically powered by an eccentric-weight vibration generator or an eccentric-driven connecting rod. The oscillations convey the refuse through the incinerator furnace from one grate section to the next, in the same manner as an oscillating or vibrating conveyor. A similar, but nonuniform, oscillation cycle is used for the impact stoker or grate. The grate sections are moved either forward or in reverse and then released for spring return against a stop (impact). The grate surfaces are generally of the same type as used in the reciprocating grate elements, and, indeed, might be considered as a variation of the reciprocating type of grate. Although oscillating and impact grates have been used in fuel burning furnaces and incinerators for many years, there are fewer of them than the other types of grates thus far described.

(d) Traveling Grate. The traveling grate is widely used in continuous flow incinerator furnaces. There are many installations in the United States, in Europe, and in Japan. The traveling grate has been in use for many years in coal fired furnaces and, as in most grate systems, has been adapted for use in municipal incinerators. There are two types of traveling grate stokers: the chain grate and the bar grate. Both convey the refuse fuel from the gravity feed chute through the incinerator furnace to the ash residue discharge, much as a conveyor belt.

Because the traveling grate stoker does not stoke or mix the fuel bed as it conveys, incinerator traveling grate stokers are often cascaded in two, three, and even four or more units. With two units in cascade formation, the first portion under the feed chute and continuing through the ignition zone is often called the feeding or drying stoker. The discharge from this first

portion of the traveling grate spills and tumbles onto a second unit, thereby
mixing, breaking, and redistributing the fuel bed on the second stage. With
three or more stages, this spillover and redistribution is repeated. Since
the thickness of the refuse fuel bed decreases as the combustion process
continues, the linear conveying speed of successive traveling grate units is
adjusted to accommodate the rate of combustion and to provide more complete burnout after the combustion rate has subsided (decreased linear speed).
Attempts have been made to install auxiliary mixing devices over the fuel
bed on the traveling grate. However, these have not been successful, mainly
because of entaglements with scrap wire and other debris contained in the
refuse. The best method for stoking or mixing with the traveling grate is to
use several grates so that stoking and redistribution is accomplished as the
burning refuse is spilled from one section to the succeeding section.

(e) <u>Drum Grate.</u> The drum grate, or roller grate (sometimes called the
barrel-type grate) was developed by the city of Düsseldorf, West Germany,
to counteract the high cost of multiple traveling grates needed to obtain
better burnout. Each rotating drum or cylinder represents a minimum
length of a traveling grate section, thereby providing a maximum number of
tumbling zones in the fuel bed. This rotating grate system consists of six
to eight or more rotating grate cylinders, 1.5 m in diameter, mounted on a
horizontal axis in close proximity and at a downward inclination of about
30°. The slow rotation of the drums creates a strong mixing action in the
refuse between successive drums, thus loosening and opening bundles of
paper or plastic bags. The rotating cylinders extend over the width of the
furnace (width of the grate) and each cylinder is equipped with its own variable speed drive to control the movement of the burning refuse through the
incinerator. Thus, the rotating drum grate provides better control of refuse
stoking and residence time during combustion than the traveling grate with
consequent improvements in burnout.

f. Incinerator Furnace Enclosures

The furnace enclosure provides a controlled environment for the combustion process in the incinerator system. Without the furnace enclosure,
the combustion process would, in effect be "open burning." If such burning
is covered with a metal shield for protection from wind and direct radiation
to the atmosphere, we have the conical or "teepee" burner which has been
used for municipal refuse and wood wastes. However, the conical unit has
the technical limitation that it must be operated with 300 to 600% excess combustion air to prevent catastrophic heat damage to the metal structure and
enclosure. Materials for conventional incinerator furnace enclosures include:

Nonmetals: refractories, e.g., firebrick walls and roofs

Metal: plate, tubes, pipe, etc.

Refractory covered metal: castable or firebrick refractory lining
or coatings, 3 to 25 cm thick

Either water or air may be used to cool these enclosure materials. Cooling
water can be either contained in tubes or pipes or uncontained in the form of
a film on the surface of the external metal plate. Air cooling can be employed with forced convection; with forced air jets impinging on the surface;
or by radiation and natural convection to the ambient environment. Warmed
cooling air can be used as preheated air for combustion or can be ducted
for building heat.

(1) Refractory Enclosures

(a) <u>General</u>. There are at least three types of refractory enclosures:
gravity walls with sprung arch roofs; suspended walls and suspended roofs;
and refractory linings supported directly by the metal shell. Gravity walls
consist of refractories laid in courses, one above the other, supported on
a foundation floor or other support, and are vertical (or almost vertical).
The sprung arch used with gravity walls is a portion of a circular arch of
refractory shapes supported by compression of the side walls through skew-
backs retained with structural steel members held together with tie rods.
Suspended refractories are supported on structural steel either directly
with clamps attached to each piece of refractory, or with intermediate
refractory pieces keyed or cast into the refractory roof being supported,
with the refractory supporting member held with a metal clamp to the structural steel. Refractory linings are installed in a circular flue or duct and
held like staves in a barrel, or if cast refractory is used, it is supported
with metal hooks or dowels fastened to the shell by bolting or welding.
Refractory walls can also be cast over pipes or steam tubes, thus protecting
the metal tubes against corrosion or flame impingement, and at the same
time cooling the refractory surface, with the tubes, in turn, cooled with hot
water or steam.

A furnace enclosure infrequently used for municipal incineration is the
rotary kiln. This is a near horizontal slowly rotating cylinder which may
be lined with castable refractory or with large brick shapes called "kiln
blocks." It is also possible to use an unlined rotary kiln cooled externally
with air or water, or additionally cooled with a water film on the inside of
the cylinder. This latter concept has only been applied in industrial incineration.

(b) <u>Refractories for Small Multiple-Chamber Incinerators</u>. The minimum
specifications for refractory materials used for lining the exterior walls of
multiple-chamber incinerators [127] are the following:

1. For waste of 2800 to 4500 kcal/kg Higher Heating Value use

Firebrick: high heat duty, Pyrometric Cone Equivalent (PCE) not
less than 32.5

Castable refractory: not less than 1920 kg/m^3 (120 lb/ft^3), PCE not less than 17

2. Wood, sawdust, and other high temperature service:

Firebrick: superduty, PCE not less than 34

Plastic: PCE not less than 34; density not less than 2040 kg/m^3 (130 lb/ft^3)

Minimum refractory thickness for lining exterior walls (including arches) of incinerators burning all classes of refuse are the following:

1. Up to and including 160 kg/hr (350 lb/hr) capacity: castable refractory or plastic: 10 cm (4 in.); firebrick: 11 cm (4.5 in.)

2. Above 160 kg/hr (350 lb/hr) capacity: all refractories: 23 cm (9 in.)

Stacks should be lined with refractory material with a minimum service temperature of 1100°C (2000°F). In low-capacity units the minimum lining thickness should be 6.5 cm (2.5 in.); in units larger than 160 kg/hr (350 lb/hr), 11 cm (4.5 in.).

Doors should be lined with refractory material with a minimum service temperature of 1550°C (2800°F). Units smaller than 45 kg/hr (100 lb/hr) should have door linings of 5-cm (2-in.) minimum thickness. In the size range of 45 to 160 kg/hr the linings should be increased to 7.5 cm (3 in.). In units with capacities from 160 to 450 kg/hr (350 to 1000 lb/hr) the doors should be lined with 10 cm (4 in.) of refractory. In units of 450 kg/hr (1000 lb/hr) and more, linings should be 15 cm (6 in.).

The thickness of refractory lining and insulation in the floor of a multiple-chamber incinerator depends primarily on its physical location. Incinerators that are installed on their own concrete foundations outside of buildings should have 7.5 cm (2.5 in.) of firebrick lining backed by a minimum of 3.8 cm (1.5 in.) of 1100°C (2000°F) insulating material. Incinerators of the semiportable type should have sufficient air space provided beneath the incinerator so that no damage to the pad will result. When incinerators are installed within buildings, it is extremely important that provision be made to prevent damage to floors and walls of the building. Such damage can be prevented by providing air passages beneath the incinerator and adjacent to the building walls to prevent excessive heat from actually reaching the structure. If an air space beneath the incinerator is impractical, then additional insulation should be provided.

For incinerators with capacities of up to 225 kg/hr (500 lb/hr), 11 cm (4.5 in.) of firebrick and 6.5 cm (2.5 in.) of insulation should be provided on the floor of the mixing and final combustion chambers. For incinerators with capacities of 225 to 900 kg/hr (500-2000 lb/hr), 11 cm (4.5 in.) of firebrick backed by 10 cm (4 in.) of insulation should be provided.

Units in high temperature service should be provided with an insulating air space of 7.5 to 10 cm (3-4 in.) between the interior refractory and the exterior steel. This will reduce the temperature of the refractory and extend its life. Adequate openings above and below the incinerator should be furnished for air to enter and exit freely. In some cases forced circulation of the air in this space may be required. Some incinerator designs utilize forced preheated air as secondary combustion air.

The minimum thickness of interior refractory dividing walls generally follows that required for the exterior walls. The bridge wall, with its internal secondary air distribution channels, requires greater thickness. The minimum width of refractory material between the secondary air channel and the ignition or charging chamber should never be less than 6.5 cm (2.5 in.) for very small units; 11 cm (4.5 in.) for units up to 110 kg/hr (250 lb/hr); and 23 cm (9 in.) for larger units.

(c) Refractories for Large Incinerators. The choice of refractory type and the details of installation for large municipal incinerators are still evolving. Table 40 indicates the more common choices for placement in the different regions of the incinerator.

In general, suspended construction is preferred; to reduce the initial cost and to reduce maintenance expense.

(2) Other Enclosure-Related Design Considerations

(a) Air Inleakage. Refractory walls and roofs can permit air to leak in or furnace gases to leak out depending on the draft conditions in the furnace, unless airtight metal casings are used. Such infiltration is particularly common with suspended refractory construction. It can be a benefit in providing air cooling of the wall or roof material, with inleakage of air adding to that required for combustion. In many installations, however, the inleakage of air can be excessive to the point that furnace temperatures are reduced, combustion of the furnace gases is quenched, and air pollution control systems are overloaded.

(b) Shape Factors. The physical shape of the furnace enclosure is a factor in controlling the incineration process. For example, there is a tendency for hot combustion gases to rise to the top of the chamber. If the furnace outlet is at the top of the enclosure, the cooler gases from the discharge portion of the grate will leave the furnace below the layer of hotter combustion gases, often without mixing. Thus a large fraction of the combustion air is not used and the flue gases may contain excess air as well as unburned combustible. This problem is reduced somewhat with chamber designs which exhaust the gases from the incinerator furnace either at the top near the refuse feed or at the lower portions of the furnace near the ash discharge.

The installation of overfire air jets, refractory baffles, or bridge walls will also aid mixing.

(c) <u>Heat Release Rate</u>. The volumetric heat release rate characterizes the combustion intensity and wall temperature level in the furnace enclosure. Although designs vary, most refractory furnaces fall within the range from 130,000 to 225,000 kcal hr^{-1} m^{-3}, with an average of 180,000. This compares unfavorably with stoker fired coal boilers which operate at 260,000 to 310,000 kcal hr^{-1} m^{-3}. Comparative heat release rates are shown in Table 41.

(d) <u>Secondary Combustion Chambers</u>. Many design specifications require a separate secondary combustion chamber with the incinerator installation. While the primary combustion chamber may be defined as the ignition, volatilization, and burning zone above the incinerator grates, the secondary combustion chamber or zone may be either a separate downstream chamber or an additional furnace volume downstream of the grate area. In either case, the secondary combustion zone provides suitable residence time (usually 1-2 sec) for completion of the gas phase combustion reactions. This secondary zone should be located downstream of the point at which additional air is mixed with the furnace gases to supply oxygen for the completion of the combustion. The temperature of the secondary combustion chamber should be maintained above 800 to 900°C to insure complete oxidation of smoke (carbon and tar particles), hydrocarbon vapors, and combustible particulate and gases (such as carbon monoxide). A tertiary chamber is sometimes added downstream of the secondary chamber to provide additional combustion time, to provide for flue gas cooling and settling of fly ash, or simply to provide a flue gas connection from the secondary chamber to the breeching and stack.

(e) <u>Slagging and Clinkering</u>. In design and operation of the incinerator, care should be given to avoiding sidewall, roof, or grate temperatures where the refuse ash becomes tacky. Under such conditions, slag will build on walls and roof or massive clinkers can form on the refuse bed. Table 42 indicates the effect of temperature and ambient atmosphere on the fusion characteristics of coal and refuse residues.

g. Combustion Air

For incineration, the term combustion air usually includes underfire air, overfire air, and secondary air. However, these terminologies apply to incinerators employing grates to support and/or convey the burning refuse through the incinerator. While combustion air for nongrate systems will be discussed later, underfire air for grate systems is defined as the air supplied upward through the grates and beneath the burning refuse. Underfire

TABLE 40

Suggested Refractory Selection for Incinerators[a]

Incinerator part	Temperature range (°C)	Abrasion	Slagging	Mechanical shock	Spalling	Fly ash adherence	Recommended refractory
Charging gate	20–1425	Severe, very important	Slight	Severe	Severe	None	Superduty
Furnace walls, grate to 48 in. above	20–1425	Severe	Severe, very important	Severe	Severe	None	Silicon carbide or superduty
Furnace walls, upper portion	20–1425	Slight	Severe	Moderate	Severe	None	Superduty
Stoking doors	20–1425	Severe, very important	Severe	Severe	Severe	None	Superduty
Furnace ceiling	370–1425	Slight	Moderate	Slight	Severe	Moderate	Superduty
Flue to combustion chamber	650–1425	Slight	Severe, very important	None	Moderate	Moderate	Silicon carbide or superduty

Combustion chamber walls	650–1425	Slight	Moderate	None	Moderate	Moderate	Superduty or 1st quality
Combustion chamber ceiling	650–1425	Slight	Moderate	None	Moderate	Moderate	Superduty or 1st quality
Breeching walls	650–2400	Slight	Slight	None	Moderate	Moderate	Superduty or 1st quality
Breeching ceiling	650–2400	Slight	Slight	None	Moderate	Moderate	Superduty or 1st quality
Subsidence chamber walls	650–870	Slight	Slight	None	Slight	Moderate	Medium duty or 1st quality
Subsidence chamber ceiling	650–870	Slight	Slight	None	Slight	Moderate	Medium duty or 1st quality
Stack	260–540	Slight	None	None	Slight	Slight	Medium duty or 1st quality

[a]From Ref. 129.
[b]Air cooling is often used with subsequent discharge of the warmed air as overfire air.

TABLE 41

Heat Release Rate for Various Fuels and Firing Conditions

Stoker	Fuel	Grate heat release (10^3 kcal hr^{-1} m^{-2})	Volumetric heat release (10^6 kcal hr^{-1} m^{-3})
Single retort (underfeed)	Bituminous coal	540	
Multiple retort (underfeed)	Bituminous coal	810	
Traveling (cross-feed)	Bituminous coal	810	0.26–0.31
	Municipal refuse	810	0.18
Spreader (overfeed)			
on stationary grate	Bituminous coal	950–1080	
	Lignite, bagasse	1220–1630	
on dumping grate	Bituminous coal	1020–1290	
	Lignite, bagasse	1360–1760	
on oscillating grate	Bituminous coal	1500	
	Lignite, bagasse	1760–1900	

on traveling grate			
Bituminous coal		1700–1970	N/A
Lignite, bagasse		2300–2700	N/A
Pulverized coal:			
dry bottom	0.13–0.20	N/A	N/A
wet bottom	0.60–4	N/A	N/A
Oil			
normal pressure	0.18–1	N/A	N/A
increased pressure	10	N/A	N/A
gas turbine	100	N/A	N/A
Gas		N/A	N/A
Nuclear reactors	0.18–0.25	N/A	N/A
gas cooled	0.43–0.86	N/A	N/A
Pressurized water reactor	17–160+	N/A	N/A
Boiling water reactor	17–26	N/A	N/A
Liquid metal cooled	170	N/A	N/A

TABLE 42

Ash Fusion Temperature Ranges of Refuse and Coal[a]

Ash Source	Reducing Atmosphere (°C)	Oxidizing Atmosphere (°C)
Refuse[b]		
initial deformation	1030–1130	1110–1150
softening (H = W)	1200–1300	1240–1325
softening (H = 1/2 W)	1210–1310	1255–1345
fluid	1315–1405	1360–1480
Coal[c]		
initial deformation	1060–1100	1105–1245
softening (H = W)	1080–1205	1160–1345
softening (H = 1/2 W)	1195–1215	1240–1355
fluid	1230–1425	1310–1430

[a]From Ref. 130.
[b]From three samples of St. Louis refuse, with magnetic metals removed.
[c]From three samples of Union Electric Company coals.

air is required to cool the grates (to maintain their structural integrity and to avoid oxidative corrosion of the grate metal). Underfire air is also supplied to furnish oxygen for the combustion reaction; it may be insufficient for complete combustion, yet sufficient to release enough heat to pyrolyze and/or gasify the refuse and remove the volatile components. It is common practice to supply a quantity of underfire air theoretically sufficient for complete combustion of the refuse on the grates. However, poor air distribution vis-a-vis the distribution of air demand, the high relative rate of gasification and pyrolysis reactions, and poor gas mixing over the bed requires additional air to be supplied as overfire or secondary air for combustion, mixing, and for dilution of the gases to maintain temperatures below 980 to 1100°C. Underfire air is required in the later stage of the batch process (or further downstream in the continuous process) for oxidation of the remaining fixed carbon in the ash residue to provide complete burnout.

Overfire air may be defined as that air admitted above the burning bed of refuse on the grate. It is usually admitted either in low or high velocity jets to mix the combustible gases rising from the burning refuse with combustion air. Secondary air, added for temperature control, may be admitted through high velocity jets in the side walls and roof of the furnace enclosure, either near the upstream end of the primary furnace or at the transition between the primary and secondary furnace. Also, secondary air can be added at low velocity through a slot or small openings in the bridge wall separating the primary chamber from the secondary chamber. In the latter case, mixing is dependent more on the shape of the chamber and changes in direction of the main gas stream than on the energy carried by the air jets.

Centrifugal fans with electric motor drives are used to supply combustion air. Since the fans for high velocity combustion air must operate at higher tip speeds and therefore with considerable noise, the fans should be located where their noise will not be a nuisance. In general, sound absorbing devices should be considered on these and other high-speed fans.

Low velocity combustion air can also be admitted through louvers, doors, or other openings as a result of the negative pressure or draft within the furnace chamber. The quantity of combustion air admitted through such openings can be controlled by dampers or by the door opening.

Combustion air for suspension burning, without main grates, requires a different consideration from combustion air for a grate system. In suspension burning within a waterwall furnace or within a refractory furnace, the air that conveys the shredded refuse into the chamber may be half or all of the theoretical air required for combustion; however, it must be sufficient to convey and inject the shredded refuse into the furnace. Good results have been obtained when air is added tangentially to the furnace chamber to create a cyclonic action, with the burning mass in the center of the rotating cyclone and the air injection surrounding the cyclonic flame. If the suspension burning system includes an auxiliary grate at the bottom of the furnace chamber for completing the burnout of the ash residue, a small amount of underfire air is desirable through this grate. Although high-speed centrifugal fans in single or multiple stage can be used to supply the air for injection of the shredded refuse for suspension burning, positive displacement blowers can also be used for injection air as well as for high velocity tangential air jets to induce mixing within the cyclone of flame in the furnace.

h. Ash Removal

After complete incineration of the refuse, the ash residue drops into an ash chamber or chute from the end of the grate or kiln. Siftings that have fallen through the grates (which may have been either partially or completely burned) and collected fly ash also may be conveyed to this ash chamber. The ash may be discharged directly into a container or onto suitable conveyors for disposal, or into water for quenching and cooling. The ash residue is then removed from the water with a drag conveyor, pusher conveyor, or

other means. Rubber belt conveyors have been used in Europe for some
time on water-quenched ash residue, and there are now a few installations
in the United States. Pneumatic and hydraulic handling could perhaps also
be used, as in the handling of the ash from the combustion of coal after it
has been burned under steam boilers. Alternatively, the ash residue can
be handled manually from the ash hopper by discharging into a dump truck
or other suitable container and transported to the residue disposal site or
serviced to byproduct recovery systems.

To prevent inleakage of air or outleakage of furnace gases at the point
where the gas is removed, an air seal is desirable. Dry mechanical seals
and seals made by covering the ash receptable or container have been used
to control air leakage. With wet removal of the ash, a wet or hydraulic
(water) seal is utilized or a combination of a wet and mechanical seal is
used. The ash residue usually includes abrasive and corrosive materials
and miscellaneous slag and metal. Therefore, to avoid malfunctions, the
effects of wire, coat hangers, cans, metal hoops, and occasional large and
heavy pieces of wood or metal, must be carefully considered in the design
of equipment which handles the residue.

In connection with the disposal of the ash, a number of installations
include recovery of salvageable materials before the remaining ash is conveyed to the dump or disposal site. Such recovery systems include rotary
screens (trommels), vibrating screens, crushing or shredding devices, as
well as ballistic and magnetic separation equipment. Through the use of
these devices, ferrous metals can be removed for salvage as scrap; medium
ash solids, such as clinker particles, portions of fused glass, or particles
of shattered glass, can be removed for use as fill material or for use in
surfacing and construction of alleys and secondary streets; the fine ash
which has no salvage value is transported by truck or by means of a belt
conveyor or a hydraulic system to the dump site for disposal.

The oversize noncombustibles in the ash residue from the incinerator
furnace are customarily transported directly to the residue disposal site.
In a few instances, oversize noncombustibles are passed through crushing
or breaker rolls which flatten larger metal pieces and crush nonmetal
pieces. The material that has been reduced in size may be passed over a
magnetic separator to remove iron for scrap salvage.

i. Flue Gas Conditioning

Flue gas conditioning is defined as the cooling of the flue gas after it
has left the combustion zone to permit discharge to mechanical equipment
such as dry air pollution control devices and fans or a stack. In general,
cooling to 230 to 370°C is necessary if the gas is discharged to mechanical
equipment, while cooling to 470 to 590°C is adequate for discharge to a
refractory-lined stack.

Both wet and dry methods are used for cooling (or tempering) incinerator flue gas streams. The sections below discuss the technical and economic features of several embodiments of these methods.

(1) Cooling by Water Evaporation

In wet methods, water is introduced into the hot gas stream and evaporation occurs. The degree of cooling is controlled by (1) the amount and droplet size of the water which is added to the gas, and (2) the residence time of the gas in the water atmosphere. Presently, two types of wet cooling are used: wet bottom method and dry bottom method.

(a) Wet Bottom Methods. The wet bottom, which is most commonly used, involves the flow of large quantities of water (much more than is required for cooling the flue gas). The water is supplied by coarse sprays operated at relatively low pressures. The excess water falls to the bottom of the cooling zone and is rejected or recycled. Although not efficient, some removal of particulate occurs by droplet impingement and gravity settling in the spray chamber.

The equipment used in this system consists of several banks of sprays, each with several nozzles with relatively large openings (over 0.5 cm). These spray banks are normally located in the flue leading to the stack or air pollution control equipment. Pumps are not normally required, since line pressure is adequate for satisfactory operation. However, pumps are required if the water is recycled. The bottom of the flue must be maintained watertight to prevent leakage.

The system is generally controlled by measuring the gas temperature downstream of the sprays and modulating water flow, either manually or automatically. Generally, spare spray banks are provided which can be pressed into service quickly in response to sudden temperature increases or plugging of the spray nozzles.

The advantages of this system are that it is relatively simple, reliable, and inexpensive to design and install. The gas is humidified during cooling, which may be somewhat advantageous for some types of air pollution control equipment (electrostatic precipitators) and disadvantageous for others (fabric filters). Also, there is a reduction in total gas volume during cooling, as shown in Fig. 38. This figure shows the change in volume of flue gas as a function of inlet gas temperature when the gas is being cooled to 260° C by means of air dilution, water evaporation, and a convection boiler.

A disadvantage of the wet bottom system is that much more water (greater than 100% excess) is used than is necessary for cooling the gas. The excess water is acidified in use and is contaminated with particulate and dissolved solids. Therefore, it normally requires treatment before it can be discharged from the plant. In some installations the excess water is

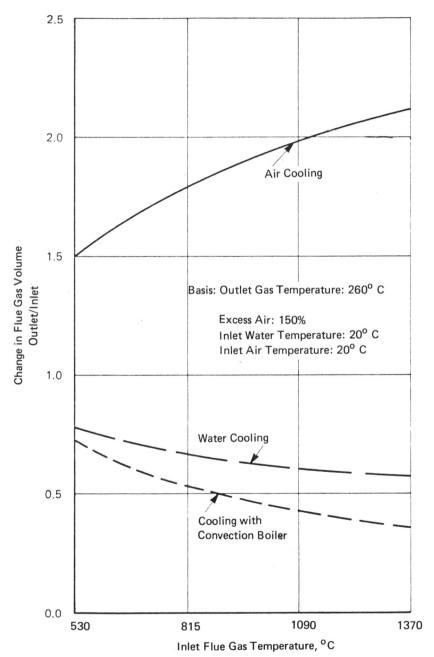

FIG. 38. Change in flue-gas volume during cooling to 260°C (to 315°C for boiler). (From Ref. 21.)

recycled to the process, and in these cases severe problems of erosion, corrosion, and plugging are encountered in the nozzles, piping, and recirculation pumps. As a result, excessive maintenance costs and equipment down-time are common. Removing suspended solids using clarifiers, hydraulic cyclones, and settling ponds is helpful, but problems with corrosion, plugging, and erosion have not been eliminated. In some plants, the clarifiers and ponds are sources of odor. Also, ponds require considerable land area and may result in groundwater pollution. Corrosion problems result from the very low pH (less than 2.0) which occurs when recirculation systems are used and the build-up of soluble salts in the recirculated water. These problems can be controlled to some extent by adding chemicals such as lime or soda ash to the contaminated water. However, such systems have significant operating costs and do not completely alleviate corrosion problems.

Another disadvantage of the wet bottom method is that flue gas leaving the spray chamber may carry entrained water droplets or wet particulate matter. These moist particles can cause operating problems with the air pollution control and fan equipment due to fly ash adherence and accumulation. Also, corrosion of unprotected metal parts downstream or build-up of hardened deposits of solids in the flues and fan housing can occur.

(b) Dry Bottom Methods. In the dry bottom method, only enough water is added to cool the gas to the desired temperature, and the system is designed and operated to assure complete evaporation. In this system, a conditioning tower 10 to 30 m high is required and fine, high pressure spray nozzles are used. Booster pumps are necessary to raise the water pressure to assure fine automization; water pressures from 6 to 36 atm are common in such systems. Alternatively, atomization is effected using compressed air or steam. Control is usually accomplished with a temperature controller measuring the outlet flue gas temperature and modulating the flow of water to the conditioning tower sprays. Since high pressures are required, only a small degree of flow rate modulation by pressure variation is practical. If the flow is too low, the pressure drop across the nozzles will fall, and poor atomization will result; if the flow is too great, the pressure drop across the nozzle will increase and limit flow. As a result, additional spray heads are put into or taken out of service as required to maintain uniform outlet temperatures. Air atomization techniques are preferred to obtain the required turn-down.

Water droplet size is very important in the design of a dry bottom conditioning tower; droplets that are too large will not completely evaporate and, in addition to the problems of water carryover described above, will cause a wet bottom in the conditioning tower. Droplets which are too fine will lead to high power usage. A balance must be made among droplet size, gas residence time in the tower, and power usage. Equation (150) presents a formula developed by Hardison, of U.O.P. Air Correction Division [131] for estimating the evaporation time of water droplets:

$$t = \frac{r_d}{0.123\ (T - T_d)} \tag{150}$$

where

t = residence time in seconds

r_d = droplet radius in micrometers

T = temperature of gas (°C)

T_d = temperature of droplet (°C)

A second and more complex method of estimating (in a step-wise fashion) the evaporation time of droplets (up to about 600 μm) uses the graph shown in Fig. 39 where

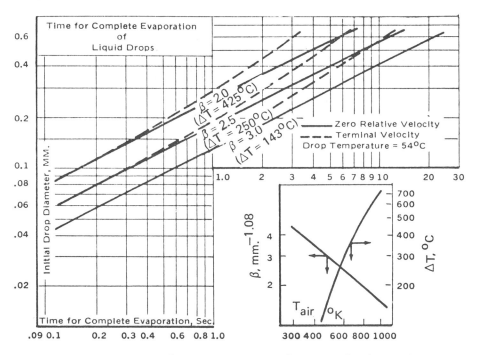

FIG. 39. Theoretical evaporation times for water droplets in hot gas streams (from Ref. 132).

T_a = gas temperature (K)

$\Delta T = T_a - 54$ (°C)

μ = gas viscosity at T_a (g cm^{-1} sec^{-1})

$\beta = 0.071 \, (\mu^2 T_a)^{-0.36}$

If the drop temperature varies significantly from 54°C, the evaporation time may be corrected by a factor $(T_a - 54)/(T_a - T_{drop})$. It is suggested that users of this latter method consult the original paper [132].

Although both of these methods are useful in estimating the residence time required for evaporation, care must be taken in selecting a nozzle which provides uniform size droplets rather than droplets with a wide size distribution, since the largest droplet will dictate the length of the conditioning tower.

The advantages of the dry bottom system are that it minimizes water consumption, eliminates water pollution problems, produces a "dry" effluent gas (free from entrained water), and reduces the volume of the flue gas as shown in Fig. 37. The gas is humidified during cooling, which may or may not be desirable. The disadvantages of the system are that it is expensive to design and install, power consumption is high, control is somewhat complex, and the small orifices of the atomizing nozzles make them susceptible to plugging. The nozzle plugging can be minimized by filtration of the water, proper draining of the nozzles when they are not in use, and maintenance of sufficient flow through the nozzles to keep them cool. A recent installation in New York City used a spray head in which a portion of the water flow was recycled, thus providing additional nozzle cooling. The effectiveness of this approach is not known at the time of this writing. Keeping the nozzles cool and draining them is essential to prevent scale formation in the nozzles due to water hardness; this scale is the principal cause of plugging. In order to minimize the risk of outages, the nozzles should be easily accessible and removable during operations, and additional nozzles should be provided for emergency use.

The dry bottom system is more costly than the wet bottom system because of the need for water filtration and the maintenance required for the high pressure pumps, nozzles, and control systems. Proper materials of construction (such as ceramic wear surfaces) are essential for good nozzle life and to minimize corrosion. Power consumption can be minimized by proper nozzle selection and, with increased investment costs, by installing larger conditioning towers (providing longer residence time) which permit less fine atomization.

(2) Cooling by Air Dilution

Dilution with air is the simplest and most reliable method for flue gas cooling. Only a damper for air control is necessary for a system with adequate draft. Although air dilution has these advantages, large quantities of air are required for air dilution cooling, increasing the demands placed upon the air pollution control equipment and the induced draft fan and stack. The increase in volume caused by air cooling is shown in Fig. 37. Such large increases in volume significantly increase the capital and operating costs of the equipment which follows the point at which dilution takes place. For these reasons, air dilution should not be used in new incinerator installations.

(3) Cooling by Heat Withdrawal

The second method of dry conditioning uses a convection boiler in which heat is removed from the flue gas by the generation of steam. The equipment consists of a convection tube waste heat boiler, economizer, and all of the auxiliary equipment required, such as boiler feed water pumps, steam drums, and water treatment facilities. In addition, an air or water condenser (the latter served by a nearby natural cooling water source or cooling tower) may be necessary to condense the steam when the steam demand is less than the generation rate. These boilers would have controls similar to conventional boilers and would require an additional fulltime operator—a licensed stationary engineer.

The advantages of this system are that heat is recovered and that the shrinkage in flue gas is greater than with any other method discussed, as shown in Fig. 37. No water is added to the system during cooling, which may or may not be desirable. In some cases, the steam generated in the boiler can be sold and produce income, or the steam can be used in the plant to induce mixing in the incinerator furnace gases (steam jets) or for space heating.

The disadvantages of this method are that the system is expensive to design and install, the boiler installation is complex to operate and requires an experienced licensed operator, corrosion and erosion problems with the boiler tubes have not been completely resolved, sticking of fly ash on the boiler tubes can occur, and reliable markets for the steam must be found.

(4) Steam Plumes

In the incineration process, water is introduced into the flue gas from the evaporation of refuse moisture and as a result of the combustion process. Water may also be added during conditioning of the gases and by wet scrubbing. As a result, the flue gas leaving the stack can contain significant amounts of water vapor which will condense, under certain atmospheric conditions, and an opaque stack effluent (steam plume) will result. Although steam plumes neither cause nor indicate air pollution (often quite the contrary), public reaction is frequently negative and vocal. While these plumes will appear white and not unpleasing in bright sunlight, on cloudy days and at twilight, they may seem dark and "dirty."

Formation of a steam plume depends upon the discharge temperature and moisture content of the flue gases, the ambient temperature, and the amount of dilution which occurs at the point of flue gas discharge. The processes which occur upon discharge of the stack gases into the ambient are as follows:

1. The gases rise from the stack following the flow pattern of a free turbulent nonisothermal jet (sometimes in crossflow due to winds). One characteristic of this flow pattern is that little mixing of the jet fluid with the ambient occurs for a distance of about one stack diameter. As a result, the plume from superheated vapors may "float" above the top of the stack.

2. The stack gases then begin to mix with cooler ambient air. As this occurs, the temperature and absolute humidity of the mixed gas falls, about in proportion to the dilution ratio (kilograms of stack gas per kilogram of ambient fluid). Thus, the state of the gas mixture, as shown on a temperature-humidity (psychrometric) chart, moves downward and to the left (see Fig. 40). The curve describing the gas state differs from a straight line (between the stack discharge and ambient temperature and humidity point) only as a result of differences in relative specific heat between the flue gas and the ambient air. This effect is often neglected, however, as it acts, in the analysis which follows, to yield a conservative estimate.

3. The mixed gas state thus moves from stack conditions toward cooler and less moist conditions. If, in the course of this change, the mixed gas conditions cross the saturation curve (Fig. 40), a steam plume will form. The gaseous part of the plume then "moves" down the saturation curve while liquid water, in the form of small droplets, is released. The degree of the resultant plume opacity is related to the amount of water to be condensed, the dilution process (which affects cooling rate and droplet number and size distribution), and the ambient temperature and humidity.

4. As dilution proceeds, the condition of the gaseous part of the plume drops from the saturation curve, and the liquid droplets evaporate. At this point, the plume has "dissipated," and any residual opacity is due to suspended particulate or aerosols. Typically, however, the high dilution at this distance from the stack effectively hides what may have been an intolerable plume opacity caused by high fly ash loading.

In the case where typical municipal incinerator flue gas is evaporatively cooled from 900 to 260°C, the treated gas will contain approximately 0.36 kg of water vapor per kilogram of dry air. This moisture includes water formed by refuse combustion in addition to that evaporated in cooling the flue gases. Under these conditions, a line drawn on a psychrometric chart

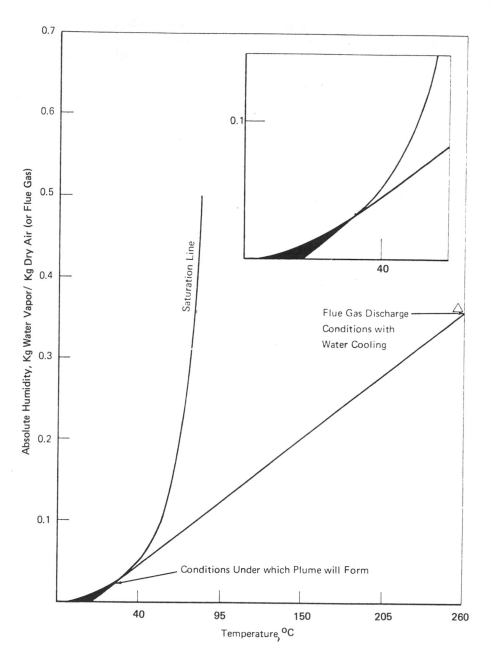

FIG. 40. Psychrometric chart illustrating steam plume formation (from Ref. 21).

(see Fig. 40) starting at 260°C, 0.36 humidity, and tangent to the saturation line, delineates the set of atmospheric conditions under which a steam plume will form. The shaded area on Fig. 40 shows these conditions. For this flue gas, a plume will form any time the ambient temperature falls below 4°C. The plume will also form at higher ambient temperatures, depending upon the absolute humidity of the ambient (falling within the shaded area).

If the flue gas is cooled to lower temperatures, say to 90 to 105°C in a wet scrubber, or if the ambient relative humidity is high, plumes will form at higher ambient temperatures. High cross-winds, providing more rapid dilution, may shorten or almost eliminate the plume. Reheating of flue gas, though very costly, will reduce the intensity or frequency of plume formation.

Since the costs of reheating will be excessive in regions of frequent low ambient temperature, it is likely that steam plumes (though possibly seasonal) will always be associated with incinerators equipped with wet scrubbers and with plants using evaporative cooling for flue gas conditioning. A number of authors have suggested that steam plumes may be reduced by subcooling the flue gases with a secondary scrubber to condense a portion of the water vapor before discharge, although this still requires a modicum of reheating. In fact, however, the heat removal required to condense this water must be compensated for in a cooling tower system (or equivalent). Such treatment, therefore, only moves the steam plume from the stack to the adjacent cooling tower. References 133 and 134 further discuss the steam plume problem.

j. Environmental Pollution Control

An incinerator is probably of greatest concern to a municipality because of the fear of the pollution of the contiguous environment. The principal concerns are air and water pollution. However, an incinerator can also create undesirable noise and cause the surrounding area to be unattractive because of litter and other forms of trash which quickly disfigure an incinerator site where good housekeeping is not regarded as a fundamental plant responsibility.

(1) Air Pollution

The most noticeable forms of air pollution are fly ash, smoke, odors (from the stack as well as other areas), noxious gases, and dust. All emanate from an incinerator at times.

(a) Composition of the Flue Gases. If combustion of the volatile fraction of the refuse is complete, the composition of the flue gas will be principally nitrogen, oxygen, and carbon dioxide. There will be small amounts of sulfur

oxides, nitrogen oxides, and traces of mineral acids (principally hydrochloric acid, which will result from the combustion of halogenated plastics,
particularly polyvinyl chloride). Normally, the concentration of sulfur
oxides, nitrogen oxides, and mineral acids will be low enough so that they
will not cause significant air pollution. It is doubtful if it is necessary to
treat flue gases to remove these materials. If combustion of the volatiles
is not complete, the flue gases will contain significant amounts of carbon
monoxide and other uncombusted or partly combusted organic materials.
The first indication of the presence of these materials in high concentrations
will be the appearance of black smoke from the incinerator stack, which
may be followed by the detection of objectionable odors.

    The presence of such unburned or partially burned materials is unnec-
essary and is caused by the poor operation of the incinerator. Their emissions should be controlled by the proper operation of the incinerator rather
than the installation of control devices. Complete combustion can be assured
by operating the incinerator at the proper temperatures (from 750 to 1000°C);
by providing sufficient air for combustion; by providing sufficient residence
time for the combustion process to occur; and by inducing (either by gas
passage design or overfire air jets) sufficient turbulence in the combustion
space to mix the combustible gases and aerosols with the necessary air.

    Such residence time and some mixing is usually provided for by ducting
the flue gases to a secondary combustion chamber. Although it is not essential that such a chamber be provided, it is necessary to provide sufficient
volume and vigorous induced mixing in the furnace to assure that the combustion process is completed. Few single-chamber incinerators meet this
need.

(b) Particulate Matter. Particulate matter (characterized by flue gas weight
loading), generally referred to as fly ash, is generated in the combustion
process and must be removed from the effluent gases. The amount of particulate matter which is generated is somewhat dependent upon the design
and operation of the incinerator. If the combustion process is not complete,
a sooty fly ash will result. The best way to control emissions of the latter
type is operation at temperatures sufficiently high to assure complete combustion of these materials.

    Studies have been made [21] which indicate that there is a correlation
between the amount of fly ash entrained in the effluent gases and the distribution and amount of overfire and underfire air and the type of grate employed.
Proper operation will assure that large amounts of fly ash do not become
entrained in the gas stream because of improper air distribution.

    There will be, however, no matter how carefully the incinerator is
operated, particulate matter entrained in the effluent gases. The extent to
which the particulate matter is removed from the gases depends upon the

type of emission control equipment which is used and the way it is operated and maintained. If abnormal amounts of particulate are being emitted, it may be that the incinerator is being operated improperly. This happens, for example, when combustion is quenched by large amounts of air admitted to the incinerator in an uncontrolled manner, such as occurs in batch feed incinerators. Also, the emission control equipment may not be operating properly. Problems which could occur in the operation of air pollution control equipment include: the plugging of spray nozzles caused by either water hardness or large amounts of solid in recirculated liquid streams, the plugging of cyclones caused by sticky particulate or condensation, and failures in the electrical systems of a precipitator.

In a properly designed and operated incinerator, equipped with appropriate air pollution control equipment, the standards established by most states and the federal government can be met.

Although the flue gases from incinerators contain a number of pollutants, air pollution control equipment installed on these units are primarily directed at the problem of particulate removal. For this purpose, a number of devices are in use, ranging in particulate removal efficiency from 5 to 15% to upwards of 95%. In light of current (1975) federal particulate emission standards (0.08 grains/ft^3 at 50% excess air), control efficiencies in excess of 95% are generally required.

Settling or expansion chambers have been used in the breeching and flue gas ducts, and many of the older installations have employed refractory baffles across the breechings extending downward from the roof or upward from the bottom of the breeching to require the flue gases to pass under and over such baffles. In some instances, a coarse spray of water is directed into the flue gases and toward the baffles with most of the water falling to the floor of the chamber without vaporization. The wet floor and baffles improve particulate removal by preventing reentrainment of settled ash into the flue gas stream. At best, however, such systems only attain a control efficiency of 20 to 35%, far below modern requirements.

Mechanical collectors are usually "cyclones" in which the flue gas is rotated within the confines of a cylinder after entering tangentially at the periphery. The flue gas then leaves through an axial outlet. Solid particulate concentrates on the inside of the cylindrical wall (as a result of centrifugal force), and solids are discharged at the lower end and opposite to the cleaned gas outlet. Listed below in order of decreasing air pollution control effectiveness (and pressure drop) are three general types of such cyclones. The maximum efficiency to be expected with such units is 60 to 80%.

1. A multiple cyclone with many small-diameter (less than 30 cm) cyclone units installed in a tube sheet

2. A multiple cyclone system of larger diameter (over 45 cm) installed in clusters with flue gas manifolded to the inlets of the individual cyclones and the outlet manifolded into a single duct

3. Single or double cyclone units of larger diameter (over 1 m) with a single or split flue duct at the inlet and outlet

Other devices used for particulate removal from flue gases include scrubbers, which may be open spray chambers, packed chambers, and most importantly, high pressure drop Venturi scrubbers. Fabric filters can, conceptually, be used in incinerator applications, but results to date have been disappointing: although collection efficiency is high (>99%), maintenance and operating problems have plagued the units. Such filter materials are mainly those of high temperature fabrics, such as silicone-treated glass fiber cloth, arranged in bags or tubes.

A common, although only moderately efficient (say, 30-40%), type of air pollution control concept for cleaning incinerator flue gases is the use of vertical staggered baffles which may be employed in multiple stages in a group, or with groups of baffles in each stage. These baffles are commonly wetted with water. Although most of the staggered baffles which are wetted with a water spray are constructed with refractory firebrick, a few are installed with corrosion resisting metal baffles, such as special stainless steels.

The electrostatic precipitator is currently receiving the greatest attention for particulate removal from incinerator flue gases. At the time of this writing, there are over seven electrostatic precipitators in the United States with operating experience. Also, a large number of European incinerators have utilized electrostatic precipitator gas cleaning for many years. These European installations were originally employed on combination power plants and incinerator furnaces, although recently there have been a few electrostatic precipitators installed on steam boiler plants in Europe which are exclusively fueled with municipal refuse. Electrostatic precipitators used for cleaning incinerator flue gases are the vertical multiple-plate type. Reported efficiencies exceed 95%, and the resulting effluent quality exceeds federal regulations. A summary of expected average control efficiencies of various air pollution control systems is given in Table 43.

(c) Smoke. Smoke (characterized by flue gas opacity) may either be ash material, which is completely burned but is very finely divided, or an unburned or partly burned combustible material (tar aerosol). If the smoke is an ash material, control requires the use of very high efficiency emission control equipment. It may be lessened to some extent through improved operation by adjusting the air distribution in the primary furnace.

If the smoke is combustible, it can best be controlled by improved combustion efficiency. Longer residence time, better air distribution, gas mixing, and high operating temperatures will eliminate such pollution.

(d) <u>Odor</u>. In addition to odors which emanate from the stack because of poor combustion, there are odors in the pit and in the residue handling area, particularly if water is used for quenching and sluicing the grate siftings. Odors may also emanate from the residue disposal landfill.

The odors from the stack can be best controlled by assuring complete combustion of the gases. This is done by proper air distribution, elevated temperatures, and proper residence time at these temperatures. Such control is much more effective and less costly than using equipment such as scrubbers for odor removal.

Odors in the pit can be controlled by using masking deodorants, minimizing the time the refuse is stored in the pit, planning the withdrawal of material from the pit, and periodic cleaning. The deodorants can be applied by using a spray network permanently installed in the pit. However, such a system should only be used in conjunction with a good cleaning program. The crane operator should establish a systematic plan for withdrawing refuse from the pit to insure that no refuse stays in the pit too long (as by working from one end of the pit to the other). Cleaning should be done once a week by washing the pit down. This can be done without completely emptying the pit by scheduling the washing of only portions of the pit at one time.

Odors in water systems used for quenching the ash and sluicing the grate siftings can only be accomplished by a very active cleaning program which involves changing the water at least once a week and through washing of the equipment. Wet ash handling systems are, at best, messy and difficult to keep clean, particularly if burnout is not complete.

(e) <u>Dust</u>. Dust is emitted from the pit area when trucks unload and when the crane is operating, from the residue handling area, and from the fly ash handling area. Dust raised during unloading can be controlled by enclosing the tipping area. Also, sprays can be used to moisten the refuse in the pit to minimize dust during the feeding. Dust in the ash handling area can be controlled by using sprays to moisten the ash and using closed transfer and hopper systems to minimize dust leakage. Some systems of these types are currently being used and have been very effective.

(2) Water Pollution

Water is used and immediately discharged from most incinerators. Water is used to wash down the pit and ash handling systems, to quench hot residue and fly ash, for dust control, and for gas quenching and scrubbing. Water which is rejected from the incinerator will most likely be polluted in one way or another.

The washwater used throughout the plant and the water which drains from the pit and ash handling systems will have high biological and chemical oxygen demands, and contain suspended and dissolved solids, both organic and inorganic [135,136]. Since the quantity of this water will be relatively

TABLE 43

Average Control Efficiency of Air Pollution Control (APC) Systems[a]

APC type	APC system removal efficiency (wt %)								
	Mineral particulate	Combustible particulate[b]	Carbon monoxide	Nitrogen oxides	Hydro- carbons	Sulfur oxides	Hydrogen chloride	Polynuclear Hydro- carbons[c]	Volatile metals[d]
None (flue settling only)	20	2	0	0	0	0	0	10	2
Dry expansion chamber	20	2	0	0	0	0	0	10	0
Wet bottom expansion chamber	33	4	0	7	0	0	10	22	4
Spray chamber	40	5	0	25	0	0.1	40	40	5

Device									
Wetted wall chamber	35	7	0	25	0	0.1	40	40	7
Wetted, close-spaced baffles	50	10	0	30	0	0.5	50	85	10
Mechanical cyclone (dry)	70	30	0	0	0	0	0	35	0
Medium-energy wet scrubber	90	80	0	65	0	1.5	95	95	80
Electrostatic precipitator	99	90	0	0	0	0	0	60	90
Fabric filter	99.9	99	0	0	0	0	0	67	99

[a] From Ref. 26.
[b] Assumed primarily < 5 μm.
[c] Assumed two-thirds condensed on particulate, one-third as vapor.
[d] Assumed primarily a fume < 5 μm.

small, it may be discharged into a sanitary sewer or septic system. It should not be discharged to storm drains or streams without treatment.

The water rejected from quenching and scrubbing systems may be quite large in quantity and very hot. It will also be high in suspended and dissolved solids and have a low pH. The solids will be both organic and inorganic. The biological and chemical oxygen demands may also be higher than desirable. In some designs, the scrubbing water is settled to remove suspended solids and recycled; in others, it is discharged after use. Recycling has often proved to be troublesome because of the plugging of lines and spray nozzles with solids and scale.

Water which is discharged should be sent to a well-designed settling pond to remove solids and permit cooling. If it is to be discharged to a stream, secondary treatment consisting of pH adjustment and possibly digestion to reduce biological and chemical oxygen demands may be required. The rate of discharge may have to be controlled to prevent the build-up of soluble solids in the stream.

If water is used only for quenching, systems could be designed to obtain 100% evaporation and thereby eliminate any disposal problem. Wet scrubbers which permit recirculation without plugging are also highly desirable.

(3) Noise Pollution

The noise created by the operation of an incinerator can be a disagreeable nuisance. Many large trucks are constantly delivering and removing material from the plant. Also, large fans, pumps, and other equipment often operate 24 hr/day. There are a number of approaches to control truck noise, including use of proper exhaust mufflers and enforcement of low-noise operating procedures. Also, the noise from the equipment in an incinerator can be controlled by use of enclosures, or silencers, where appropriate (particularly on the large fans).

k. Incinerator Stacks

Several types of stacks or chimneys are used to discharge incinerator flue gases into the ambient atmosphere. Stub stacks are usually fabricated of steel and extend a minimum distance upward from the discharge of an induced draft fan. Stacks taller than about five diameters and less than 30 m high are often referred to as "short" stacks. These are constructed either of unlined or refractory lined steel plate, or entirely of refractory and structural brick. "Tall" stacks are constructed of the same materials as the short stacks above, and are used to provide greater draft than that resulting from the shorter stack, and to obtain more effective diffusion of the flue gas effluent. Some metal stacks are made with a double wall with

an air space between the metal sheets. This double wall provides an
insulating air pocket to prevent condensation on the inside of the stack and
thus avoid corrosion of the metal. The draft or negative pressure within a
modern incinerator is commonly provided with an induced draft fan. The
need for a fan (as opposed to natural draft) is occasioned by the high pressure drop of modern air pollution control devices, the depressed stack
temperatures occurring when wet scrubbers are used, and the control advantages realized with mechanical draft systems.

The value of stack gas temperatures are higher than the ambient, draft
will be produced by the buoyancy of the hot gases. The theoretical natural
draft may be estimated by

$$P_{ND} = 0.0342 P_a L_s \left(\frac{1}{T_0} - \frac{1}{T_s} \right) \tag{151}$$

where

P_{ND} = theoretical draft, in atmospheres

P_a = barometric (ambient) pressure, in atmospheres

L_s = height of stack above the breeching, in meters

T_0 = ambient temperature (K)

T_s = temperature of stack gases (K)

The value used for T_s should be an average value taking into account
heat losses and (for tall stacks) the temperature drop on expansion [see
Eq. (1)] as the atmospheric pressure declines with increasing altitude. As
the stack gases pass up the stack, frictional losses reduce the fraction of
the theoretical draft which is available. Frictional losses may be estimated
from the following:

$$F_s = \frac{2.94 \times 10^{-5} L_s (\bar{u})^2 (\text{Perimeter of stack})}{T_s (\text{Cross-sectional area of stack})} \tag{152}$$

where

F_s = friction loss, in atmospheres

$\bar{u}$ = mean stack gas velocity (m/sec); perimeter and area in meters
and square meters, respectively.

In addition, some of the draft is lost to the velocity head in the rising gases.
This "loss," referred to as the "expansion loss" may be estimated by

$$F_e = \frac{1.764 \times 10^{-4}(\bar{u})^2}{T_s} \tag{153}$$

where F_e is the expansion loss in atmospheres.

Typical natural draft stacks operate with a gas velocity of about 10 m/ sec.

2. Incineration Systems for Solids, Sludges, Liquids, and Gases

Incinerators for sludges, combustible gases, commercial and industrial wastes generally differ greatly from those used for mixed municipal solid waste.

Systems are usually selected which are especially adapted to burn a particular waste type rather than to be a general purpose trash disposal unit.

System designs are generally more complex, and are heavily instrumented for control and monitoring.

Systems are available to incinerate pastes, sludges, liquids, and gases, as well as loose, bulky solids.

Environmental control concerns often extend to gaseous species (sulfur dioxide, hydrochloric acid, etc.) and may include the chemistry (e.g., heavy metal content) of residues.

In the sections which follow, several examples are given for incineration systems in common use for a wide variety of waste types. They are categorized by the nature of the wastes (solid, sludge, liquid, and gas). It should be recognized, however, that with adaption often more than one waste type can be burned in a given system (though perhaps with some difficulty).

Each of the incinerators includes the basic elements of feeding means, refractory or protected metal enclosure, waste support such as a stationary or moving grate or refractory hearth, a combustion air supply and flue gas exhaust system, a means of withdrawing solid residues, and, if needed, air pollution control devices. Many include manual or automatic control systems. It is beyond the scope of this book to describe, in detail, the design concepts, preferred choices for materials of construction, demonstrated performance levels, and operating strategy for each of the incinerator types. Many such presentations would fill several volumes if, in fact, agreement among "experts" could be found.

a. Incinerators for Solids

The solids of concern here include "trash" (waste usually containing a large fraction of paper and cardboard), relatively dry food waste, waste wood, agricultural residues (bagasse, corn cobs) and brush, organic containing bulky wastes (automobile bodies, rail cars), etc. Many of these solids include substantial amounts of noncombustible (wire, steel banding, drums, nails, as well as "ash"). "Standard" or reference waste compositions and incinerator classifications have been developed by the Incinerator Institute of America (Tables 44 and 45) for the specification of on-site incinerator systems.

Although not delved into in depth here, the materials handling problems of waste feeding, entanglement during passage through the burning zone, and ash removal are critical, though more to efficient and reliable operation than to combustion performance. These potential problems, as well as the risk of explosion of containerized volatile combustible, etc., is abetted by the difficulty in identifying and/or removing "troublesome" materials prior to feeding.

(1) Multiple Chamber (Hearth or Fixed Grate)

Multiple-chamber incinerators are among the most simple, usually consisting of a refractory-lined furnace chamber with a hearth floor or simple fixed grate. In small trash burning units (50-250 kg/hr), combustion air is supplied by natural draft. Large units, handling several tons per hour of bulky wastes, are often equipped with forced draft. For even minimally acceptable burnout of smoke, a secondary chamber is necessary. Auxiliary burners are often needed to maintain secondary chamber temperatures above 650 to 700°C during start-up or shut-down.

The design of large units, suitable for the combustion of 1000 to 2500 kg/hr of tree stumps, demolition waste, sofas, chairs, mattresses, etc. (heat of combustion approximately 400 kcal/kg), can be based on the following [117]:

Furnace floor hearth area corresponds to 75 kg hr^{-1} m^{-2} (300,000 kcal hr^{-1} m^{-2}).

Furnace height 3.75 m (to assure clearance during charging).

Furnace length 6 to 8 m.

Furnace width 2.5 to 5 m.

Secondary chamber volume 65% of furnace.

TABLE 44
Classification of Wastes[a]

Classification of wastes[b]	Principal components	Approximate composition (% by weight)	Moisture content (%)	Incombustible solids (%)	kcal/kg of refuse as fired	kcal of auxiliary fuel per kg of waste to be included in combustion calculations	Recommended minimum kcal/hr burner input per pound waste
Trash, type 0	Highly combustible waste, paper, wood, cardboard, cartons, including up to 10% treated papers, plastic or rubber scraps; commercial and industrial sources	Trash, 100	10	5	4700	0	0
Rubbish,[b] type 1	Combustible waste, paper, cartons, rags, wood scraps, combustible floor sweepings; domestic, commercial, and industrial sources	Rubbish, 80 Garbage, 20	25	10	3600	0	0
Refuse,[b] type 2	Rubbish and garbage; residential sources	Rubbish, 50 Garbage, 50	50	7	2400	0	800

Garbage,[b] type 3	Animal and vegetable wastes, restaurants, hotels, markets, institutional, commercial, and club sources	Garbage, 65 Rubbish, 35	70	5	1400	800	1700
Animal solids and organic wastes, type 4	Carcasses, organs, solid organic wastes, hospital, laboratory, abattoirs, animal pounds, and similar sources	Animal and human tissue, 100	85	5	550	1670	4500 (2800 primary) (1700 secondary)
Gaseous, liquid or semiliquid wastes, type 5	Industrial process wastes	Variable	Dependent on predominant components	Variable according to wastes survey	Variable according to wastes survey	Variable according to wastes survey	Variable according to wastes survey
Semisolid and solid wastes, type 6	Combustibles requiring hearth, retort, or grate burning equipment	Variable	Dependent on predominant components	Variable according to wastes survey	Variable according to wastes survey	Variable according to wastes survey	Variable according to wastes survey

[a]From Ref. 137.
[b]The figures on moisture content, ash, and kcal as fired have been determined by analysis of many samples. They are recommended for use in computing heat release, burning rate, velocity, and other details of incinerator designs. Any design based on these calculations can accommodate minor variations.

TABLE 45

Classification of Incinerators[a]

Class I:	Portable, packaged, completely assembled, direct-fed incinerators, having not over 0.14 m^3 storage capacity, or 11 kg/hr burning rate, suitable for Type 2 waste.
Class IA:	Portable, packaged, or job assembled, direct-fed incinerators 0.14 m^3 to 0.42 m^3 primary chamber volume; or a burning rate of 11 kg/hr up to, but not including, 45 kg/hr of Type 0, Type 1, or Type 2 waste; or a burning rate of 11 kg/hr up to, but not including, 35 kg/hr of Type 3 waste.
Class II:	Flue-fed, single-chamber incinerators with more than 0.19 m^2 burning area, for Type 2 waste. This type of incinerator is served by one vertical flue functioning both as a chute for charging waste and to carry the products of combustion to atmosphere. This type of incinerator has been installed in apartment houses or multiple dwellings.
Class IIA:	Chute-fed multiple chamber incinerators, for apartment buildings with more than 0.19 m^2 burning area, suitable for Type 1 or Type 2 waste. (Not recommended for industrial installations.) This type of incinerator is served by a vertical chute for charging wastes from two or more floors above the incinerator and a separate flue for carrying the products of combustion to atmosphere.
Class III:	Direct-fed incinerators with a burning rate of 45 kg/hr and over, suitable for Type 0, Type 1 or Type 2 waste.
Class IV:	Direct-fed incinerators with a burning rate of 35 kg/hr or over, suitable for Type 3 waste.
Class V:	Municipal incinerators suitable for Type 0, Type 1, Type 2, or Type 3 wastes, or a combination of all four wastes, and are rated in tons per hour or tons per 24 hr.
Class VI:	Crematory and pathological incinerators, suitable for Type 4 wastes.
Class VII:	Incinerators designed for specific byproduct wastes, Type 5 or Type 6.

[a]From Ref. 137.

Forced air supply corresponds to 150% excess, supplied 25% through 7 cm diameter floor ports at 30 m/sec; 75% through 8 cm diameter sidewall ports. Sidewall port nozzle velocity selected to give 10 m/sec at furnace midplane. Assume 500 m/min inflow velocity through charging door clearances.

Area of flue ("flame port") between furnace and secondary designed to flow at 500 m/min for 175% excess air at mean firing rate; 650 m/min between secondary chamber and outlet breeching to the air pollution control system.

Such furnaces have been constructed and tested [117, 118, 121] and been found satisfactory in this service for an operating cycle of 7 hr of charging at 25 min intervals, and a burn-down through the next 17 hr. Similar designs can be developed for car bodies, drum cleaning, and other bulky waste incineration, although consideration should be given to the heat release rate relative to the above. Auxiliary burners may be added to the furnace to hold temperatures at or above 550°C to assure acceptably high burning rates. The heat input requirements may be estimated using the basic heat and material balances, heat loss, and infiltration estimates presented above.

For commercial or light industrial waste disposal in the capacity range from 20 to 1000 kg/hr, a well-established design rational has been developed. The design, based on an extensive empirical investigation undertaken by the Air Pollution Control District of Los Angeles County over the years 1949 to 1956, has been reduced to a series of nomographs, tables, and graphs relating construction parameters to burning rate (Ref. 138 and 139 and reported in Ref. 127).

The Los Angeles designs are meant to be applied to wastes with a heating value from 2400 to 3600 kcal/kg (typically paper, rags, foliage, some garbage, etc.). The unit designs are predicated upon the use of natural draft to draw in air under the fire (about 10%), over the fire (about 70%), and into the secondary (mixing) chamber (about 20%). Typically, the total air ranges from 100 to 300% excess.

The Los Angeles design methods are applied to two configurations: the retort and the in-line. The retort style derives its name from the return flow of flue gases through the secondary combustion chambers which sit in a side by side arrangement with the ignition (primary) chamber. The in-line style places the ignition (furnace), mixing and combustion (secondary combustion) chambers in a line. The retort design has the advantages of compactness and structural economy and is more efficient than the in-line design at capacities to 350 kg/hr. Above 450 kg/hr, the in-line is superior.

(2) Multiple Chamber (Moving Grate)

These units are similar in design and operation to the municipal incinerators described above. Clearly, their general capacity range 50 to 300 tons/day per furnace restricts their economical use to the larger industrial complexes.

In many cases such units are equipped with burners for waste oils, pitches, solvents, etc., as well as cage dryers for drying wastewater treatment plant sludges prior to incineration. Some plants have provisions for feeding drums to be fire-cleaned.

Since the waste supporting surfaces of most moving grate systems are perforated or designed with clearances to admit underfire air, care must be taken to control the nature of the feed to minimize the occurrence of serious undergrate fires. For example, drums of flammable solvents, tars, and other thermoplastic materials which flow when heated should not be charged. Steam or fog-type water spray fire control systems and/or, more commonly, a water-sluiced siftings removal system should be considered if the probability of such occurrences is high or if incinerator outages are critical to maintaining overall plant facility operation.

(3) Starved Air

In the 1960s, a number of manufacturers, recognizing the increasing emphasis on smoke abatement, began producing incinerators which limited (or starved) the combustion air supply. This holds flue gas temperatures high and minimizes the need and/or fuel consumption for afterburner devices.

The starved air units consist of a cylindrical or elliptical cross-section primary chamber incorporating underfire air slots in the hearth-type floor. Overfire air is also supplied. All combustion air is provided by forced draft fans. In many such incinerators, the proportion of underfire to overfire air is regulated from a temperature sensor (thermocouple) in the exhaust flue (higher exit temperatures increase overfire air and decrease underfire air).

Most units incorporate a gas or oil fired afterburner which is energized whenever the exit gas temperature falls below the set point temperature (usually 650-750°C). The afterburner is either mounted in a separate secondary combustion chamber or in a refractory-lined stack.

Starved air incinerators are available in capacities from 100 to 1000 kg/hr, and if properly operated and maintained can meet most federal, state, and local air pollution codes when fed with typical office and plant trash (principally paper, cardboard, and wood).

(4) Open Pit Type

The disposal of scrap wood, pallets, brush, and other readily combustible, low-ash wastes is often a problem. Open burning is prohibited in many communities and, because of its slowness and inefficiency, is both costly (labor costs) and presents safety hazards.

Monroe [140] developed a simple refractory-lined pit incinerator which has shown utility in these applications. His concept uses perforated pipes in the floor of a rectangular in-ground or above-ground pit to supply forced underfire air. Jet nozzles, directed such as to "bounce" off a side-wall onto the fire, are mounted along one side of the pit.

Similar incinerator designs have been used for the disposal of heavy pitch and tar residues, but with variable success and generally great concern on the part of air pollution regulatory agencies.

(5) Conical (Tepee) Type

The conical or tepee incinerator, as its name implies, is a large, perforated metal cone. Waste (usually brush and other low density or bulky wood waste) is charged to the unit on a batch basis. Air is typically supplied by natural draft but, in some units, a refractory lined floor is installed with holes for forced underfire air. Excess air levels, by design of the sidewall perforations, exceed 400 to 600% to maintain the structural integrity of the steel shell and to minimize oxidative metal wastage. Such units are little more than a bonfire in a box and offer few advantages when stringent air pollution regulations must be met. They have enjoyed some utility in large scale land clearing or lumbering operations.

b. Incinerators for Sludges

It has been recognized that the massive increases in secondary municipal and industrial wastewater treatment forecast for the period 1970 to 1985 will generate veritable mountains of sludge at treatment plants throughout the United States. In addition, a wide variety of chemical and biochemical manufacturing operations (petroleum refining, coke manufacture, etc.) generate large quantities of sludge wastes. These wastes vary in solids content over a very wide range (2-100%) but, if incineration is to be considered, are generally dewatered to at least 5 to 20% solids.

Because (most) sludges can be pumped, the problems of storing and feeding found for solid wastes are greatly simplified (although odor can be a severe problem in prolonged storage of biological sludges). Conversely, the flow characteristics and small ash particle sizes of sludges make the use of a grate-type support during burning unacceptable. For this reason, sludge is burned on a hearth or in suspension.

A second and important characteristic of high-moisture sludge is the "protective" action of the free water, greatly slowing the combustion rate. On introduction into a hot environment, the outer layer of sludge dries and chars. The ash and char layer provides insulation, reducing the rate of heat transfer to the interior. The high latent heat of evaporation of the water in the interior further extends the time required for complete drying and combustion. The net impact of these effects is that the sludge incinerator must provide means to continually abrade the sludge mass to wear off the ash/char layer and expose the wet interior to heat.

Three incinerator systems are commonly used to cope with these unique physical demands: the rotary kiln, the multiple-hearth, and the fluidized bed. It is worth noting that all of these systems can and are used for the incineration of solid wastes, although with the following "preparation" of the solids before entry:

For the rotary kiln, solid wastes are often preburned on a moving grate. The kiln, rotating slowly to avoid undue suspension of dust, is then a specialized "burnout grate."

For the multiple-hearth, the waste must be cleaned of metal wire and other components which will rapidly accumulate on the rabble arms which plow and convey the waste through the unit.

For the fluidized bed, the waste should be cleaned of massive inerts which would be difficult to remove from the bed. Also, the waste should be shredded to avoid rapid blow-off of larger pieces and to promote rapid burning.

(1) Kiln Systems

A rotary kiln is comprised of a cylindrical, refractory-lined steel shell, supported at two or more (but preferably only two) points. The kiln is driven by a gear mounted external to the kiln, usually near one of the support points. The kiln is sloped gently (usually, less than 3 cm/m) and rotated slowly (usually, less than 2 rpm). The internal surface of the kiln may be smooth or contain longitudinal plates or ridges (flights) to lift and spill the material, circumferential ridges (dams) to hold back material, and/or a multitude of baffles (internals) to improve the contact of the solid with the air or flue gas flowing through the kiln. Solids flow can be either concurrent or countercurrent to the gas flow.

The rate of movement of material through a kiln may be estimated by:

$$t = \frac{0.19\,L_T}{(rpm)\,DS} \tag{154}$$

where

t = the mean residence time (min)

L_T = the length of the kiln (m)

rpm = the kiln rotational velocity (revolutions per minute)

D = the kiln <u>inside</u> diameter (m)

S = the kiln slope (m/m)

The effect of dams and other internals on mean retention time has been described by Bayard [141].

In operation, sludge is fed to the higher end of the kiln. Kiln loading is usually in the range of 3 to 12% of the cross-sectional area. The time-temperature profile experienced by the sludge depends upon the mode of operation: concurrent or countercurrent.

In concurrent flow, a relatively large burner is used in the feed hood to apply heat to the incoming wet sludge. By the time a significant portion of the sludge has dried to the point where ignition can take place, the sludge is often too far from the primary burner to effect reliable ignition. A second kiln with an ignition burner would then be appropriate. The high cost of a second kiln, drive, and supports, and the high fuel consumption (note that the heat release from the burning sludge is unavailable for wet sludge drying) make this strategy unattractive.

In countercurrent flow, the burner is mounted at the discharge end of the kiln, firing against the direction of sludge movement. Entering sludge passes down the kiln, into ever-hotter regions: the drying zone, heating zone, ignition and burning zones, and, just before discharge, the ash cooling zone. As the gases pass up the kiln, they are progressively cooled, but are in contact with cooler sludge. Thus the combustion energy of the sludge and the burner are most efficiently extracted. Unfortunately, it is just this efficiency which creates a problem: low wet-end temperatures and consequent odorous off gas. In the energy conscious times of the late 1970s, the high fuel costs for odor destruction (using an afterburner) of the typically high gas volumes (corresponding to 200-300% excess air) in such operations are often unacceptable. Clearly, however, well-dewatered sludges, operations to hold wet-end temperatures as high as possible and excess air control can minimize this serious disadvantage.

(2) Multiple Hearth

The multiple hearth furnace (Fig. 41) is the most widely used wastewater treatment sludge incinerator in the United States. A typical unit consists of a vertical cylindrical shell containing from four to twelve firebrick hearths. Mounted in the center is a hollow cast iron shaft, rotating at 0.5 to 1.5 rpm, to which opposed rabble arms are attached. Both the center shaft and the rabble arms are cooled by air forced through a central "cold air tube" by a blower and returning in the annular space between the cold air tube and the outer walls: the hot air compartment.

The units range in diameter from 1.37 m to 6.55 m, and in capacity from 100 to 3600 kg/hr of dry sludge. Each hearth has an opening (drophole)

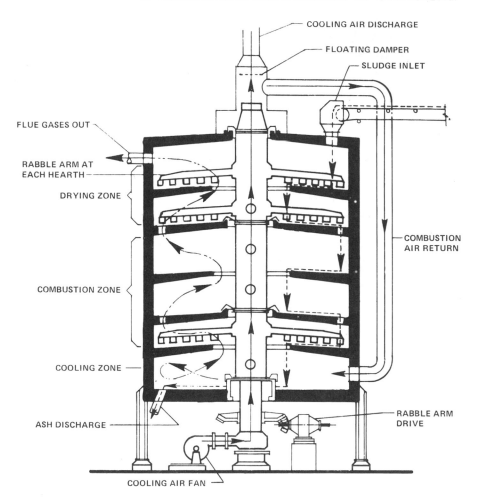

FIG. 41. Multiple-hearth sludge incinerator.

through which the sludge falls from hearth to hearth, alternating in location from along the central shaft (an in-feed hearth) to at the periphery (an out-feed hearth) such that the bottom hearth is out-feed.

The rabble arms provide mixing action and mechanical abuse to the sludge as it passes through the drying, burning or combustion zone, and cooling zone. The shape of the plows on the rabble arms both moves the sludge toward the exit opening and generates ridges to increase the effective exposed area to up to 130% of the plan area. Retention time is proportional to shaft rpm, the number of rabble arms, and the plow settings. An average of 80 min retention time is required for a 20% solids sludge.

The temperature profile in the countercurrent sludge and flue gas flows depends on the relative magnitude of the following energy terms:

Sludge character and feed rate (moisture content, dry solids heating value, and ash content).

Combustion air quantities and temperature: In some units the warmed air from the hot air compartment (typically 120-175°C) is used for combustion air. Also, each hearth has two doors which leak some air.

Fuel firing rate: burners for oil or gas are often provided on the upper and several lower hearths. Also, screenings, grease, and scum are often added to one of the lower hearths.

Heat loss from the outer shell.

In the idealized case, the temperature profile is as shown in Table 46.

In theory, the multiple-hearth incinerator can be operated without generating an odorous off-gas: little odoriferous matter is distilled until 80 to 90% of the water has been driven off (a sludge solids content of say 70%), and, at this point in the furnace, flue gas temperatures are high enough to burn out the odor. In practice, uncompensated for variations in sludge moisture and/or heat content, inattentive or untrained operators, inadequate mixing and/or residence time of odorous off-gas, and other factors occur with sufficient frequency to almost assure that odor will be a problem, at least from time to time. Protection from such problems includes the use of the top hearth as a secondary combustion chamber (with auxiliary fuel firing as needed) or installation of a separate afterburner chamber. Drophole size is also selected to maximize gas mixing by accelerating the flue gas without causing undue entrainment of particulate matter.

In the design of multiple-hearth units for municipal sludge service, the parameters listed in Table 47 are reported design or operating value for basic parameters.

TABLE 46

Idealized Temperature Profile in Multiple Hearth Furnace Burning Municipal Wastewater Treatment Sludge

	Drying zone	Burning zone	Cooling zone
Sludge	70°C	730°C+	200°C
Flue gases	425°C+	830°C+	175°C+

TABLE 47

Typical Design Parameters for Multiple-Hearth Incinerators

Parameter	Units	Low	Mean	High
Hearth area burning rate	kg dry solids hr^{-1} m^{-2}	7.2	9.8	16.2
Excess air	percent overall	20	50	80
Cooling air exit temperature	°C	95	150	195
Discharge ash temperature	°C	38	160	400
Off-gas temperature	°C	360	445	740
Sludge properties				
heat content	kcal/kg volatiles	5300	5550	7760
volatile content	% of dry solids	43.4	54.2	71.8
Total energy input (fuel plus sludge)	kcal/kg wet sludges			
	(for ~25% solids sludges)	810	1100	1355
	(for ~48% solids sludges)	1595	1730	1922
Volumetric heat release (fuel plus sludge)	kcal hr^{-1} m^{-3}		67,600	

Multiple hearth furnaces may also be used for the incineration of industrial sludges and have been used to burn mixtures of municipal sewage sludge and shredded, demetalized municipal refuse [142].

In the latter application, sludge is fed to the top hearth, prepared refuse either to the top or to an intermediate hearth. Several European plants where refuse is fed to the intermediate hearth have experienced unacceptable odor emission problems (no afterburner was installed). This problem apparently results from the variability in refuse heat content and consequent swings in the temperature of zones critical to odor destruction. Details on these operations are available [143-147].

Data from the Uzwil co-incineration plant in Switzerland [147] indicate an estimated hearth heat release of 57,000 kcal hr^{-1} m^{-2} with a hearth loading of 31.3 to 33.8 kg hr^{-1} m^{-2} and a volumetric heat release of 42,700 kcal hr^{-1} m^{-3}. The exit gas temperature was 880°C. It should be noted that in the Uzwil plant, the top two (of 12 original) hearths were removed to provide additional combustion space over the feed hearth.

Problem areas in multiple hearth operation in sludge burning service (beyond the odor problems described above) include high refractory and rabble arm maintenance and the long time required to bring the units into service. Refractory and rabble arm problems (high maintenance) arise, particularly from rapid temperature excursions (too rapid rise to temperature and high temperatures due to localized overfiring). Ideally, the furnace should be brought to temperature (or cooled down) over a 24- to 30-hr period. Severe operating problems are also presented when temperatures rise high enough to fuse the residue (clinkering). Under these conditions, large slag masses form which may build up on the rabble arms, bind rabble arm movement, or block the drop-holes. Such problems can be severe when high concentrations of inorganic salts are present in the wastes.

(3) Fluid Bed

The bluid bed was developed, originally, to provide a means to contact finely divided catalysts with high molecular weight petroleum feedstocks in a cracking process. Subsequent development has lead to a wide range of applications where the objectives include realizing rapid heat and mass transfer to the incoming feed (solid or gas); uniform temperatures, high thermal inertia (insensitivity to minor variations in feed character); and relatively short residence times (less than 5 sec).

The fluidized bed incinerator (Fig. 42) is comprised of a vertical, cylindrical, refractory-lined vessel with a perforated grid in the lower section which supports a sand bed. Consider the response of such an arrangement as the flowrate of gas upward through the bed increases:

Initially, friction produces a pressure drop which increases with velocity.

At some point the pressure drop has increased until it is equal to the sum of the bed weight per unit cross-sectional area plus the friction of the bed against the walls. Beyond this velocity the bed either lifts and moves as a piston up and out of the chamber (unlikely with loose, granular solids), or the bed expands or fluidizes so that the gas can pass without the pressure drop exceeding the bed weight. Pores and channels appear in the early stages of bed expansion.

With further increases in gas velocity, particle-to-particle spacing increases and the violence of movement fills and eliminates the gross channels. Gas and particles circulate about in the bed with some transient streams of extra-high or extra-low particulate concentration but, in general, a high degree of bed uniformity.

The size of solids which may be fluidized varies from less than 1 μm to 10 cm; with the best (most uniform) operation for particles between 65 mesh and 10 μm. Gas velocities through the beds range from 0.15 to 3 m/sec (based on the empty cross-sectional area: the superficial velocity). More detailed information on fluid bed design and operational characteristics may be found in Refs. 148 to 150.

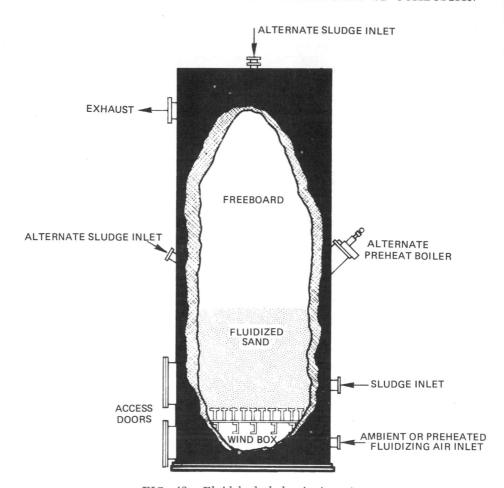

FIG. 42. Fluid-bed sludge incinerator.

Fluid bed systems are comprised of a reaction vessel containing the fluid bed, a disengaging space (freeboard), and a gas distribution plate. Solids or sludges are fed into the bed or into the freeboard and are rapidly heated by radiation and intense convection. The rapid heat and mass transfer result in high temperature uniformity, usually with not more than a 5°C temperature differential between any two parts of the bed.

In sludge burning, excess air levels are maintained at 20 to 25% to minimize fuel costs but assure complete oxidation of all volatile solids in the sludge cake. Note that with sludge feeding into the freeboard, some sludge drying (odor emission) takes place outside the bed, thus mandating the higher than bare minimum excess air levels that can be tolerated when sludge is introduced into the intensely mixed fluidized zone.

In order to conserve energy, the hot (say, 800°C) exhaust gases from the fluid bed may be passed through an air preheater. Preheating the air to 550°C, for example, can make a substantial reduction in operating cost.

As sludge burns out in the bed, the finer ash particles are swept from the bed. As a consequence, highly efficient air pollution control devices should be used with these units. The coarser particles in the sludge residue accumulate in the bed. On balance, however, there is often a net loss in the sand bed due to abrasion and disintegration such that periodic sand addition is required.

The high sensible heat content of the sand bed provides a stable heat reservoir. When the unit is shut down, the bed will easily maintain an almost constant temperature, for example, overnight, thus permitting a rapid start-up in the morning.

Fluid beds typically have a somewhat lower capital cost than multiple-hearth units (their primary "competitor" for sewage sludge incineration) but a higher operating cost due to higher fuel and power usage.

Problem areas with fluid beds for sewage sludge or other liquid or sludge waste disposal center in two areas:

1. Control of the temperature throughout the bed and flues

2. Clinker (slag) formation

The former problem is particularly serious if the waste is fed into the free-board. If continuous or periodic ignition occurs in this area (where mixing is very imperfect) overheating of regions of the flue may occur causing slag buildup and necessitating shutdown. Clinker formation within the bed and flues can occur at any time, but especially when the feed contains a high concentration of inorganic salts which fuse to a hard clinker. Tests of the waste residues using differential thermal analysis methods [151] or by placing the residue in combustion boats and holding the mass at progressively higher temperatures in a muffle furnace will serve to screen for the conditions where clinker formation is likely. Nonetheless, changes in sludge character may alter the fusion point in the future such that clinkering must always be a matter of particular concern in the selection of fluid bed systems.

c. Incinerators for Liquids

In many industrial processes, waste liquids are produced for which disposal by incineration is necessary. Such liquids include wastewater contaminated with combustible toxic chemicals, solvents or oils for which purification costs are excessive, and heavy pitches and tars.

In many instances waste liquids are burned in larger furnaces designed primarily for solids disposal. However, several incinerator designs have been developed for liquid firing alone. These furnaces are comprised of cylindrical chambers with atomized liquid and combustion air introduced axially or tangentially. Also, the fluid bed incinerators described above may be used for liquid waste incineration.

(1) Liquid Storage

Often, liquid waste incinerators are operated only a fraction of the work week or day. Thus, tanks are required to hold the waste. The design of such containers and associated piping should include careful consideration of the following.

1. Corrosive Attack: The liquids to be stored may range widely in chemical composition. The availability of an incineration system in a plant will often result in its use (and misuse) for waste streams perhaps not envisioned during design. The materials of construction of the storage tank, piping, valves, pumps, etc., should be selected with this in mind.

2. Chemical Reactions: Particularly in complex chemical plants, a wide variety of wastes may be sent to the incinerator system. In many cases, the exact composition is unknown and the full range of possible reactions cannot be investigated. Reactions of concern include:

Exothermic reactions which liberate enough heat to cause boiling, accelerate corrosion, etc.

Polymerization reactions which could solidify or turn the tank contents into an unpumpable gel

Gas forming reactions which could cause foaming or otherwise force liquid out of the tank

Precipitation reactions which could produce unacceptably large quantities of solid sludge in the tank

Pyrophoric reactions which could result in spontaneous ignition of the tank contents

3. Phase Separation: Often various waste streams are immiscible or contain settleable solids. Upon setting, phase separation or settling can occur such that the incinerator will experience major changes in feed composition over a short time, an undesirable condition. Decanting systems, recirculating pumps and/or agitation may resolve these problems.

4. Abrasion: The presence of solid phase in the wastes can cause rapid abrasion in pumps, valves, and piping.

5. Freezing and/or Viscosity Increase: Many wastes freeze or become viscous at ambient temperatures. In such cases heating of the containers and steam or electrical tracing of piping may be necessary to avoid freeze-ups or pumping problems.

6. Vaporization: Introduction of hot wastes into the storage tank may volatilize low-boiling compounds, sometimes explosively. Proper care in waste dumping and the installation of tank vents (with flame arrestors) are appropriate countermeasures.

(2) Atomization

Atomization is the process of physically breaking up a liquid into particles. Liquid wastes should be atomized so that the combustion air can quickly surround the surface of the droplets to produce a combustible mixture. Also, fine atomization speeds the rate of vaporization of the waste, a prerequisite for ignition and combustion.

A number of methods are available to effect the atomization of liquid wastes. They vary in their relative capital and operating cost, their maximum capacity (liters/min), the proportion of combustion air to be supplied as secondary air, the range of operating rates (turndown) required, and the desired flame shape.

The minimum energy input requirements for atomization are determined by the viscosity of the waste at the atomization point. The kinematic viscosity (absolute viscosity/density) of the liquid is often used to characterize atomizer requirements. In the c.g.s. system, the units of kinematic viscosity are square centimeters per second, or "stokes." For oils and other viscous liquids, the centistoke gives numerical values in the 0 to 100 range. The more common unit is based upon a determination of the kinematic viscosity by measurement of the time t (in seconds) of efflux of a fixed volume of fluid through a short standard capillary tube. The commercial viscometers are designed to obey the empirical relationship

$$\nu = a_1 t - \frac{a_2}{t} \quad cm^2/sec \text{ (stokes)} \tag{155a}$$

or

$$t = \frac{\nu}{2 a_1} + \left[\left(\frac{\nu}{2 a_1} \right)^2 + a_2 \right]^{1/2} \tag{155b}$$

For common viscometers [4]:

Viscometer	a_1	a_2
Saybolt Universal (SSU)	0.0022	1.8
Redwood No. 1	0.0026	1.72
Redwood No. 2 (Admirality)	0.027	20
Engler	0.00147	3.74

It is also useful to recognize that, for many liquids, a plot of absolute viscosity versus absolute temperature on log-log paper is, essentially, a straight line.

Example 18. Measurements of the kinematic viscosity of a waste oil indicate 200 SSU at 99°C and 400 SSU at 83°C. The specific gravity of the oil in this temperature range is 1.02. To what temperature should the oil be heated to have a kinematic viscosity of 85 SSU, the viscosity recommended for atomization?

From Eq. (155a), the kinematic and absolute viscosities corresponding to the Saybolt Universal determinations are:

	SSU	(Centistokes)	(Centipoises)
at 99°C (372 K)	200	43.1	44.0
at 83°C (356 K)	400	87.6	89.3

The objective is to find the temperature where the oil has a viscosity of 85 SSU which corresponds to $\nu = 16.58$ or $\mu = 16.91$. Plotting μ versus absolute temperature on log-log paper gives the result 122°C.

(a) Low Pressure Air Atomization. These burners require air at from 0.035 to 0.35 atm, usually supplied by blowers. The minimum air pressure requirements (energy cost) are determined by the viscosity of the liquid to be atomized. A heavy pitch with a viscosity (heated) of 80 to 90 SSU requires air at over 0.1 atm, whereas aqueous wastes can be atomized at 0.035 atm. Usually, the waste liquid is pumped to a pressure of from 0.3 to 1.2 atm. Turn-down for these burners ranges from 3:1 up to 6:1. The air used for atomization ranges from 2.8 to 7.4 m^3/liter, with lesser air being required as the air pressure increases. The resultant flame is comparatively short as, even for a pure fuel oil, about 40% of the stoichiometric air is intimately mixed with the fuel spray as the mixture enters the furnace.

(b) <u>High Pressure Air or Steam Atomization.</u> These burners require air or steam at pressures in excess of 2 atm and often to 10 atm. Atomizing air consumption ranges from 0.6 to 1.6 m^3 air/liter of waste as the supply pressure varies over this range. Steam requirements range from 0.25 to 0.5 kg/liter with careful operation (a wasteful operator may use up to 1 kg/ liter). Waste heating to reduce viscosity is required only to the extent needed for pumpability.

In general, high pressure atomizing burners show poor turndown (3:1 or 4:1) and consume considerable compressor energy or steam (typically, for boilers, about 2% of the steam output). Since only a small fraction of the stoichiometric air or an inert gas (when steam is used) is mixed with the emerging fuel spray, flames from these burners are relatively long.

In burning high-carbon pitches and tars, the addition of steam has been shown to reduce the tendency for soot formation. This apparently results from the enhanced concentration of hydroxyl radicals which act as vigorous oxidants with the unsaturated carbon radicals which are precursors to soot.

(c) <u>Mechanical Atomization.</u> These burners atomize by forcing the liquid, under high pressure (5 to 20 atm), through a fixed orifice. The result is a conical spray into which combustion air is drawn. In its simplest embodiment, the waste is fed directly to a nozzle. With such an arrangement, turndown is limited to 2.5 to 3:1, since, for example, a 75% reduction in pressure (atomization energy) is required to reduce the flow rate by 50%. Thus atomization effectiveness (droplet fineness) drops rapidly as the burner moves off the design flowrate.

The second type of mechanical atomizing nozzle incorporates a return flow such that a much smaller change in atomization pressure is needed to effect a change in flow rate. For these atomizers, turndown can be as great as 10:1. The viscosity of fluids atomized mechanically need not be as low as that for low pressure air atomization: 150 SSU is a typical design value.

The flame from mechanical atomizing burners is usually short, bushy, and of low velocity. The half-angle of the flame can be altered somewhat by changing the atomizing nozzle but, because all the air is provided by secondary means, combustion is not as rapid as with other types of burners, and a larger combustion space is usually required.

Mechanical atomizing burners are usually applied where large peak capacities (40-4000 liters/hr) are required and where large turndown ratios are desirable. Its capital cost is higher than for other designs, but subsequent operating costs are low. In waste burning applications, consideration must be given to the problems of erosion and plugging of small orifices due to solid matter in the waste stream.

(d) <u>Rotary Cup Atomization</u>. These burners atomize by the action of a
high speed rotating conical metal cup from the outer edge of which the
waste liquid is thrown into a stream of low pressure air entering the incinerator around the cup. The rotary cup is usually attached to an extension
of the low pressure centrifugal blower shaft, and the waste liquid is delivered to the cup at low pressure through the shaft or at the side of the cup
at its inner edges.

Since the rotary cup system has little requirement for fluid pressurization, it is ideal for waste burning applications where the solids content of
the waste is high. Also, the viscosity of the waste need only be reduced to
150 to 330 SSU. Turn-down is about 5:1, and burners with capacities from
4 to 1000 liters/hr are available.

The flame shape from rotary cup burners is similar to that from
mechanical atomizing burners, but with a somewhat increased combustion
rate since a portion of the combustion air is supplied with the waste
stream.

(3) Ignition Tiles

In order to assure rapid ignition of the waste-air spray, a refractory
block or "ignition tile" is used. The tile usually consists of a conical
depression with an opening at the small end of the cone which mates to the
atomizing burner. Its objective is to facilitate lighting, maintain ignition
under all normal conditions, and confine the air introduced by the burner
so that it will be properly mixed with the vaporizing waste. Its design
affects the shape of the flame and the quantity of air which can be induced
by the burner.

When the heat content of the waste is low or when the combustion chamber temperatures are too low to secure complete combustion, a refactory
tunnel extension is often added to the ignition tile, increasing the intensity
of radiation (for waste vaporization). When wastes contain a substantial
fusible ash content, special care must be given to the type and design of
these tiles to avoid rapid fluxing losses and slag build-up.

(4) Combustion Space

The combustor heat release volume requirement ($kcal/m^3$) depends
upon the combustibility of the waste and the mean furnace temperature.
For difficult to burn wastes or low furnace temperatures, more volume is
needed than for the reverse. Typical ranges are:

Temperature (°C)	Combustion volume (kcal hr^{-1} m^{-3})
300–800	30,000–130,000
800–1100	130,000–350,000
1100–1400	350,000–500,000
1400–1650 +	500,000–900,000

Flue areas should be chosen to balance the desire to minimize the infiltration of tramp air (i.e., keep furnace pressure elevated) and yet avoid pressurization of the furnace which will inhibit the flow of needed combustion air. For systems operating such that 25 to 50% of the air is to be induced by natural draft, the total air supply approximates 20% excess, and furnace gas temperatures are about 1000°C; approximately 0.25 m^2 per million kcal/hr heat release is a typical design point. If all of the air is supplied by forced draft, one-half this flue area is typical. However, it should be recognized that a large number of variables are involved in such determinations, and a careful analysis of furnace flow dynamics is appropriate prior to setting flue dimensions.

(5) Incinerator Types

Incinerators for liquids are typically comprised of simple cylindrical refractory-lined chambers.

(a) Axial or Side Fired Nonswirling Type. In these units, the burner is mounted either on the axis or in the side-wall, firing along a radius. Such units are simple to design and construct although they are relatively inefficient in the use of combustion volume. For these systems, the lower combustion volume heat release rate parameters are appropriate. In essence, these units simply provide a hot refractory enclosure in which to burn wastes and collect the flue gases for pollution control.

In the design, special care should be given to realizing good turbulence levels to assure that a large fraction of the combustion volume is utilized. To this end, high pressure secondary air jets are appropriate. Care should also be given to evaluate the probable flame length (to avoid flame impingement).

(b) Vortex Type. To increase the efficiency of utilization of combustion space, swirl burner or tangential entry designs are commonly used. In

these systems, one of two designs are commonly used: the axial swirl burner or tangential inlet cyclonic designs. The design concepts for these systems are described in Chapter 4, Section C.1.b.

d. Incinerators for Gases (Afterburners)

(1) Current Engineering Technology

A large body of knowledge, both theoretical and empirical, is available
to support and guide design and evaluation of afterburner systems. This
knowledge which concerns the heat transfer, fluid mechanics, and kinematics
of combustion phenomena, can be used to determine the design characteristics for optimum performance and thus to develop criteria against which
to evaluate existing afterburner systems.

(a) Direct Flame Afterburner Technology. Considerable experience exists
in the use of direct flame afterburners for the combustion of gaseous and
gas-borne combustible pollutants [152,153]. Many of these systems consist
of little more than a burner in a cylindrical refractory-lined chamber and
are constructed on-site by facilities engineers. Indeed, the designs of
many direct flame afterburners now on the market are derived from such
"homemade" devices.

(i) Combustible Gaseous Pollutant Control. Most applications for afterburners concern the destruction of combustible gaseous emissions. The
concentration of these pollutants is usually too low to permit self-sustaining
combustion because they are often intentionally diluted below the lower
flammability limit for safety reasons. As a result, external energy must
usually be added. The energy content equivalent to the lower limit of flammability of most gas-air mixtures is 0.46 kcal/sm^3; therefore, fuel must
be introduced into the gas stream to increase the energy potential, and
permit subsequent ignition and oxidation of the mixture. Because preheat
(sensible energy) in the pollutant stream is equivalent to chemical energy,
the flammability energy limit (kcal/sm^3) decreases as the inlet temperature
increases until it is zero at the ignition temperature.

However, operating a system at the limit of flammability (either
through sensible or chemical heat input) increases the probability that
residual partially burned combustible materials will appear in the effluent.
The presence of these materials (carbon monoxide, formaldehyde, methanol, etc.) generally indicates either inadequate mixing or quenching (by
dilution air or cooled surfaces). The combustion characteristics of such
homogeneous systems are described below.

Combustion Kinetics. In general, the combustion of hydrocarbon vapors is controlled by the mixing processes within the system, rather than

by combustion kinetics [154]. The classical results of Longwell and Weiss [155], Hottel et al. [11], and Mayer [156], on combustion in well-stirred reactors have clearly demonstrated the extremely high combustion intensities possible (10^7 to 10^9 kcal hr^{-1} m^{-3}) when mixing processes are eliminated as rate controlling steps. It has also been shown by these workers that the reactions of hydrocarbons or ketones are very fast relative to that of the carbon monoxide intermediate formed in the course of the oxidation reactions. It is also clear that all combustion reactions proceed by a free-radical mechanism and thus are susceptible to wall quenching and the action of radical stabilizing species such as NO_2 and branch chain hydrocarbons.

Nerheim and Schneider studied the burning rates of carbon monoxide and propane premixed with oxygen, hydrogen, and water vapor in various proportions over ranges of equivalence ratios and pressures [11]. Burning rates were determined from metered flow rates and analysis of reactor products. The final relationships presented by these authors are shown in Eq. (22).

The kinetic mechanism proposed by Nerheim and Schneider which fits the data for CO called for the rate-limiting step shown in Eq. (156a), equilibrium for Eqs. (156b), (156c), and (156d), and a three-body chain terminating step.

$$CO + OH = CO_2 + H \qquad\qquad (156a)$$

$$OH + H_2 = H_2O + H \qquad\qquad (156b)$$

$$H + O_2 = OH + O \qquad\qquad (156c)$$

$$O + H_2 = OH + H \qquad\qquad (156d)$$

The mechanism proposed for propane combustion involved the addition of a very fast reaction of propane to CO and H_2O at the expense of OH, O, and H. The small difference between the two correlations suggests the commanding role of CO combustion in hydrocarbon oxidation reactions.

Combustion kinetics for more complex compounds such as ethers involve more complex steps at lower temperatures. For example, the slow oxidation of diisopropyl ether at temperatures between 360 and 460°C apparently consists of the production and combustion of methyl radicals, the process being facilitated by aldehydes, particularly acetaldehyde [157]. As the temperature increases over 450°C, the thermal pyrolysis of the ether becomes of great importance in facilitating the production of radicals.

Fluid Dynamics. Through the action of free radicals in combustion processes, intense recirculation patterns near the flame front can be extremely important in augmenting the combustion rate. By such means, free radicals can be returned to the ignition areas. Thus, mixing in hydrocarbon afterburners not only promotes intimate contact of fuel and air but also returns activated species for rapid initiation of the ignition and combustion reactions. Recirculation effects can be viewed by evaluating homogeneous combustion systems in terms of plug flow and well-stirred regions. A number of excellent studies by Hottel, Essenhigh, and others illustrate such techniques.

An example of a system that uses recirculation to enhance the burning rate is given by the swirling jet [13]. In addition to increasing the combustion intensity, the axial recirculation vortex causes burning gases to travel back toward the burner, thereby piloting the flame and increasing its stability. The appearance of an optimum swirl number for combustion systems suggests the ability of the designer to control the ratio of well-mixed and plug flow zones to maximize combustion intensity and flame stability [158].

(ii) Combustible Particulate Pollutant Control. Many industrial processes emit particulate matter. The composition and particle size distrubition of these pollutants vary widely, from the fine inorganic dusts of the mineral pyroprocessing industry to the high moisture and volatile matter, low-ash dusts of the grain milling industry and the dry low-ash furnace dusts of the channel black industry. In many cases, the recovery of product values or the relatively large dimensions of the particles suggests the application of conventional particulate control systems (cyclones, precipitators, filters, scrubbers, etc.) rather than destruction. As the particle size drops below 50 μm, however, and as the value of the material to be recovered diminishes (e.g., the aerosols from a drying oven or the soot from a solid waste incinerator), the applicability and desirability of direct flame incineration increases.

The sections below discuss the important combustion parameters which apply to the burning of such particulate materials:

Retention Time. One of the primary considerations in the design of afterburners is the retention time. In principle, retention time is a derived quantity, calculable as the sum of the time for preheat of the particulate and the gas stream and the appropriate combustion time for the particulate or gaseous species under consideration, as influenced by the dynamics of the combustion system. In practice, however, the retention time is often considered to be constant, and consideration is not given to the possibility of reducing it (and thus system cost) through manipulation of the controlling parameters [159]. This arises, in part, from the assumption that the system designer cannot greatly alter the times required for the various process steps and from the relative complexity of the analysis and computations required. In fact, the designer does have a measure of control over the combustion time of particles (through temperature and air control), the preheat time for the gas stream, and, to some extent, gas phase combustion rates. (The latter can be modified by utilization of mixing and recirculation principles; recirculation is important for incineration of hydrocarbon vapors, as discussed below.)

The time required to heat the entering gas to the furnace temperature is dependent upon the combustion intensity within the system and the incremental temperature rise required. This time (in seconds) may be computed as follows:

$$t = \frac{C^\circ_{p,av} \, \Delta T}{\dot{Q}_V} \tag{157}$$

where

$C^\circ_{p,av}$ = gas heat capacity (kcal m^{-3} $^\circ$C^{-1}) over range of ΔT

ΔT = required temperature rise ($^\circ$C)

$\dot{Q}_V$ = combustion intensity (kcal m^{-3} sec^{-1})

In low-pressure gas jet mixers, the combustion intensity can be as low as 0.009 kcal sec^{-1} m^{-3}, but in premixed mechanical burners it ranges above 4.45 kcal sec^{-1} m^{-3} [153]. Typical values for premixed high-pressure gas jet, multiple-port burners range from 1.0 to 1.4 kcal sec^{-1} m^{-3}.

Often the difficulty in obtaining high combustion intensities has been attributed to the limitations of homogeneous combustion reaction kinetics. Evidence to the contrary is provided by studies on well-stirred reactors [11, 155, 156] in which air and fuel are premixed and fed into the reactors through small holes. The resulting high-velocity (often sonic) jets promote an intense mixing of the fuel/air mixture with the products of combustion. These experiments have clearly shown that the kinetics of oxidation for a large number of fuels are so fast that volumetric heat release rates of 90 to 9000 kcal sec^{-1} m^{-3} are achievable over the 1100 to 1700°C temperature range.

Particle Heating. The initial step in the combustion of particulate matter is to raise the surface temperature of the particle to levels where oxidation reactions can occur at significant rates. In general, the time required to heat particles of the size range of interest for afterburners to combustion temperatures is small compared to the actual particle burning time. This is shown in the Nusselt number correlation for zero relative gas velocity (i.e., when the particulate is moving at the same velocity as the gas):

$$h_c = \frac{2\lambda}{d} \tag{158}$$

where

h_c = heat transfer coefficient (kcal m^{-1} hr^{-1} $^\circ$C^{-1})

λ = gas thermal conductivity (kcal hr^{-1} m^{-2} ($^\circ$C/m)$^{-1}$

d = particle diameter (m)

Evaluation of this equation at 700°C for a 50-μm particle, for example, gives an overall heat transfer coefficient of almost 2440 kcal hr^{-1} m^{-3} °C^{-1}. Under such high flux conditions, the particle temperature rises quickly to that of the ambient gas, and combustion ensues under mixed chemical reaction- and diffusion-rate control.

Particle Combustion. The design requirements for the burnout of carbonaceous particles can be determined from a consideration of the rates of oxygen diffusion to the particle surface and the chemical kinetics of carbon burning. The time for burnout for the particles can be shown to be inversely proportional to the oxygen partial pressure, to increase with particle size, and to decrease with increasing temperature [13]. The approximate burning time in seconds is given by Eq. (27).

The two terms within the large brackets in Eq. (27) represent the resistances due to chemical kinetics and diffusion, respectively. At the temperatures found in incinerators, the burning rate is usually limited by chemical kinetics. Some uncertainty exists concerning the kinetics and mechanism of carbon and soot combustion, but this does not influence the general conclusions derived from application of Eq. (27).

The combustion time (t_b in seconds) for droplets of hydrocarbon liquid of a molecular weight MW, a minimum size of 30 μm, and a velocity equal to that of the gas may be computed as follows [153]:

$$t_b = \left(\frac{29,800}{P_{O_2}} \right) MW_{(T)}^{-1.75} d^2 \tag{159}$$

where

$\quad P_{O_2}$ = the partial pressure of oxygen in the ambient (atmospheres)

$\quad\quad T$ = the furnace temperature (K)

$\quad\quad d$ = the droplet original diameter (centimeters)

(b) Catalytic Afterburner Technology. The principle underlying the catalytic afterburning of organic gases or vapors derives from the fact that, from the point of view of thermodynamics, these substances are unstable in the presence of oxygen, and their equilibrium concentrations are extremely small. With a catalytic afterburner, the combustible materials may be present in any concentration below the flammability limit. The factors which influence the combustion are temperature, pressure, oxygen concentration, catalyst selected for use, nature of the materials to be burned, and the contact of the material with the catalytic surface.

(i) Catalyst Systems. An oxidizing catalyst is used. Metals (platinum, etc.), metal oxides, semiconductors (vanadium pentoxide, etc.), and complex semiconductors (spinels such as copper chromites, manganese cobailites, cobalt manganite, etc.) are all known to promote oxidation of hydrocarbons. Since the surface catalytic reaction consists of a number of elementary acts such as the breaking and formation of bonds in the reactant molecules and electron transfer between the latter and the solid catalyst, the electronic properties of catalyst surfaces are important. Consequently, the catalytic activities of metals and semiconductors would be expected to differ due to their different electronic properties. However, under conditions of oxidation catalysis, many metals become coated with a layer of oxide, and this might be the reason why the mechanisms of hydrocarbon oxidation on metals and on semiconductors are so similar [160].

Attempts have been made to correlate the activity patterns with the electronic structure of the catalysts [161-163], the d-electron configuration of cations [164], and the heat of formation of oxides [165]. Only moderate success has been achieved, and at present a general theory of oxidation catalysis is not available. Consequently, the choice of an active oxidation catalyst is still based upon extensive empirical information or rather coarse approximations.

In catalytic afterburner practice, platinum with alloying metals is prevalent because of its high activity and the lower temperature needed to induce catalytic oxidation when compared with the other catalysts. It is either deposited upon nickel alloy ribbons and formed into filter-like mats, or deposited upon small, thin ceramic rods for the fabrication of small blocks or bricks. Other possible catalysts include copper chromite and the oxides of vanadium, copper, chromium, manganese, nickel, and cobalt.

Particularly relevant is a great deal of work done recently in connection with automobile emission control. Since automobile engines have various modes of operation (cold starts, idling, high speed, acceleration, and deceleration), the catalyst system must be effective over a wide range of exhaust temperatures, gas flow rates, and gas compositions. In contrast, the conditions encountered in stationary afterburner systems are much more steady; hence, any catalytic system suitable for automobile emission control should be effective in afterburners.

(ii) Catalytic Oxidation Kinetics. Basic data on catalytic oxidation of hydrocarbons have been available for many years. Anderson et al. [166] showed that even a fairly refractory hydrocarbon gas like methane can be oxidized completely on a precious metal/alumina catalyst at 400°C or less. In general, the higher molecular weight hydrocarbons are more easily oxidized than the lower, and hydrocarbons of a given carbon number increase in reactivity according to the following series:

Aromatics < branched paraffins < normal

Paraffins < olefinics < acetylenics

Some kinetic data are also available in the literature. Work by Caretto and Nobe [167] paid particular attention to the catalytic afterburning of some substances at low concentrations in air. They determined the burning rate of saturated and unsaturated hydrocarbons, aliphatics, aromatics, and carbon monoxide on copper oxide-alumina catalysts. The rate equations were found to be not of integral orders, and the activation energies were in the region of 15 to 27 kcal/mol of combustible substance.

In general, the course of a catalytic reaction can be conveniently considered in five steps as follows:

1. The reactants diffuse from the body of the gas onto the surface of the catalyst.

2. The reactants are adsorbed into the surface.

3. The adsorbed species interact in the surface.

4. The oxidation products are desorbed after the chemical reaction.

5. The desorbed products diffuse into the body of the gas.

Any one of these steps could be the slowest step, whose rate would determine that of the overall catalytic reaction. When the rates of several steps are comparable, they will jointly determine the rate of the overall reaction.

The catalytic afterburner usually operates in the region where diffusion rate is important. Vollheim [168], for instance, found that at temperatures up to 300°C with copper chromoxide as a catalyst and up to 270°C with palladium, the burning rate of propane is determined by the reaction rate on the catalyst surface. At higher temperatures, the effect of diffusion became increasingly noticeable.

In comparison, we may cite that the most commonly encountered hydrocarbons and combustible organic vapors require catalyst surface temperatures in the range of 245 to 400°C to initiate catalytic oxidation. Some alcohols, paint solvents, and light unsaturates may oxidize at substantially lower catalyst surface temperatures, while aromatics from the tar melting processes may require higher initiation temperatures. Hydrogen, on the other hand, will undergo catalytic oxidation at ambient temperatures. Most of the catalytic burners operate between 345 and 540°C, where diffusion rate is important. Theoretical relations between mass diffusion and chemical reaction on the catalyst surface are well developed [169,170].

Laboratory data must be used with caution, because most of the supporting studies have used granular catalyst support beds and have paid little or no attention to the fluid pressure drop. Industrial catalyst systems for fume abatement, however, have had to design with relatively open structures to minimize the fluid resistance. Since mass transfer is important, it should be expected that the geometry of these open catalyst support structures should greatly influence fluid flow, hence mass transfer behavior.

The influence of pressure on the reaction rate of catalytic combustion processes has received little attention. This oversight should be remedied because, if reaction rates increase with pressure, capital cost might be reduced by operating the afterburner combustion under moderate pressure.

(2) Afterburner Systems

Process exhaust gases containing combustible contaminants released at concentrations within or below the flammable range can, in most cases, be destroyed effectively by either furnace disposal or catalytic combustion. Properly designed, applied, operated, and serviced, either system can produce oxidation and odor reduction efficiencies exceeding 98% on hydrocarbons and organic vapors. The choice of one over the other will usually be based on initial, operating, and service costs and safety rather than on efficiency. With the thermal disposal technique, the residence temperature may vary from 510°C for naphtha vapor to 870°C for methane and somewhat higher for some aromatic hydrocarbons. The operating temperature of the catalytic afterburners is usually about 340 to 540°C. Hein [171], summarizing the economic significance of low operating temperature, showed that costs for direct flame incineration range from 150 to 600% of the cost of catalytic oxidation, depending on various factors. Furthermore, a catalytic afterburner minimizes the problem of NO_x generation during disposal. Catalytic systems are, in fact, used for the chemical reduction of these oxides [172].

(a) Direct Flame Afterburner Systems.

(i) Types. Furnaces for removing undesirable gases or particulate matter from the exhaust of a chemical or manufacturing process by direct flame incineration are, in general, either similar or identical to those used for generating heat. There are many design variations in conventional furnaces; so are there a variety of incinerator systems. The principle differences between the two types of furnaces are obviously related to the introduction of the effluent feed stream and the construction materials that are necessary to withstand special erosive or corrosive effects.

Most direct flame afterburners utilize natural gas. Since in many cases gas is employed as the primary fuel in the process to which the afterburner is applied, installation may be relatively simple, and it produces a cleaner exhaust than does burning fuel oil. Gas burner designs are conventional and may be either of the premixed or diffusion type: that is, the fuel and air are either mixed prior to entering the furnace or they are introduced separately. In some cases the (primary) air is premixed with the fuel and additional (secondary) air is introduced into the combustion chamber. The method of fuel-air injection generally defines the type and performance of the afterburner. Typical assemblies include ring, pipe, torch, immersion, tunnel, radiant flame, and static pressure burners.

Atmospheric, or low pressure, burners generally are of relatively simple design, while high pressure burner systems either require a source of high pressure gas and air or special equipment in the burner. Low pressure systems tend to produce a lower combustion intensity, resulting in larger combustion chambers and, in some cases, lower efficiencies; however, their initial cost may be less when the plant does not have high pressure gas and air supplies.

The method of introducing the effluent feed stream into the furnace depends upon its composition and the type of gas burner; however, since it generally contains a high percentage of air, it is usually fed into the system in essentially the same manner as air is introduced into conventional burners.

Other significant components of the furnace system relate to methods of (1) enhancing the mixing of fuel, air, and waste gas, (2) holding the flame in the desired position within the chamber, (3) preventing the flame from flashing back through the waste gas feed stream to its source, and (4) removing undesirable gases and particulate matter either before entering or after exiting from the furnace.

The furnaces are either constructed of high-temperature alloys or are lined with refractory materials. Alloys must be selected on the basis of design stress at maximum temperature and on the known corrosive effects of the feed stream and combustion products. Refractories offer the advantages of (1) providing insulation for reducing heat loss, (2) radiating heat back into the chamber gases and particulates, and (3) resisting erosion and corrosion. The application of refractories in furnaces is a well-developed art; thermal shock, shrinkage, spalling, and deformation characteristics have been established for a variety of refractory materials.

When economically justified, heat is recovered from the products of combustion of afterburners, either by the addition of heat exchangers in the exhaust stream or by using a conventional (boiler) furnace as the afterburner. The heat exchangers used to recover heat from the afterburner exhaust are of conventional design and include the many variations of both recuperative and regenerative systems. Among the furnaces that are modified to be used both as afterburners and as a source of heat for some other purpose are boilers, kilns, and chemical reactors.

(ii) Design and Performance Characteristics. The effectiveness of afterburners in the removal of the pollutant from the waste gas depends primarily upon the temperatures achieved within the afterburner, the mixing, and the residence time. For a given rate of throughput, higher temperatures and longer residence times provide higher levels of removal of the combustible pollutants, but require more fuel and larger combustion chambers, thus leading to higher operating and equipment costs.

In most current systems, operating temperatures are relatively high, so that the time for chemical reaction to take place is short compared to mixing times. Particles larger than 50 to 100 μm may require a relatively long residence time; however, they are generally removed by techniques other than combustion.

Since mixing of the fuel and air and of burned products with the unburned material is important to most furnaces as well as to afterburners, considerable effort has been expended in increasing combustion intensity and turbulence. Among the methods that are either being employed or investigated are insertion of the gases tangentially within the combustor, high-velocity injection of the gases, multiple-ported injection of fuel and air, baffles, recirculation of hot products into the unburned zone, and injectors that introduce swirl. In general, the design of afterburners has not received that attention applied to conventional furnaces, since the economic impact derived by improved design is low compared to that attainable with improvements in, for example, boiler furnace efficiency. In addition, the afterburner cost can be quite small compared to that for the overall equipment and operating costs of the associated chemical or manufacturing process.

As with all other processes, the demand to decrease the cost of operating afterburners will continue; however, we also expect that the need to reduce pollution from all sources will require that afterburners be more effective and their use more widespread. These factors will demand that more attention be applied to afterburner design. Obviously, except for when waste heat recovery is employed, the primary performance criteria of direct flame afterburners are considerably different from those for conventional furnaces: that is, the percentage of pollutant removed per unit of fuel is one of the principle objectives for afterburners, while for conventional furnaces it is the usable thermal energy per unit of fuel. These differences should be carefully examined and exploited in the search for improved afterburner systems.

(b) Catalytic Afterburner Systems.

(i) Types. In most cases, the chief aim in the design and operation of the catalytic afterburner is to keep the input of additional energy as low as possible. In Fig. 43a, the sole purpose of the catalytic unit is to purify the polluted air by passing it, if necessary after preheating, over the catalyst into the atmosphere. In Fig. 43b, the heat content of the purified air is used to preheat the waste gas not yet purified by means of a heat exchanger. In Fig. 43c, the heat content still remaining after this stage is utilized as heating energy for the original process emitting the waste gas. This scheme is applicable to the processes where the waste gas carries a large fraction

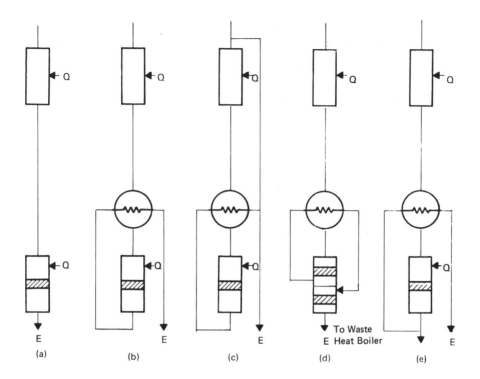

Heat Exchanger

Catalyst Bed

Q — Heat Input (Preheater)

E — Exhaust

FIG. 43. Types of catalytic afterburners: (a) straightforward catalytic afterburning of waste gas; (b) catalytic afterburner coupled with heat
exchanger; (c) catalytic afterburner coupled with heat exchanger and hot-
air recycling; (d) two-stage afterburners with heat exchanger; (e) catalytic
afterburner with cooling surfaces imbedded in the catalyst, coupled with
heat exchanger.

of the total energy consumed. In the cases where the concentration of
burnable material is very high, the two-stage afterburner scheme, as
shown in Fig. 43d, will allow the heat exchange to be operated at lower
temperatures. Further recovery of the heat content of the waste gases is
easily carried out in a subsequent waste-heat boiler. When the concentrations of organic materials in the waste gases are still higher, imbedding
cooling surfaces in the catalyst bed is preferable, since most the heat of

reaction can be removed close to the point of generation, thus keeping the maximum temperature in the reactor low and avoiding material selection problems.

(ii) Catalysts. Many substances exhibit catalytic properties, but metals in the platinum family are conventionally used because of their ability to produce the lowest ignition temperatures. Others are copper chromite and the oxides of copper, chromium, manganese, nickel, and cobalt.

In industrial use, the active component of the catalyst is usually deposited on a carrier material, as for example in the following combinations:

Active metal/metallic carrier (e.g., platinum deposited on nickel alloy ribbon)

Active oxide/oxide carrier (e.g., copper oxide deposited in alpha-alumina)

Active metal/oxide carrier (e.g., platinum/gamma-alumina)

The following are some of the important characteristics of catalysts:

Catalyst composition

Catalyst total surface area

Surface areas of the active components

Pore volume

Pore size distribution

Initial catalytic activity toward selected reaction

Compression strength

Surface area decrease upon use

Change in pore volume/pore size distribution upon use

Change in catalytic activity upon use

Change in catalyst composition upon use

The catalyst will gradually lose activity through fouling and erosion of the catalyst surface, so that occasional cleaning is required, and eventually it must be replaced. The susceptibility of catalysts to poisons varies, but the following occurrences are common.

Solid particles foul the catalyst.

Metallic vapors such as mercury and zinc deactivate the catalyst.

Phosphorus oxidizes to phosphoric acids, which deposit, on cooling, on the catalyst surface.

Sulfur containing compounds form sulfate and coat the catalyst surface.

For chlorine containing compounds, which present no problem under normal operating conditions, temporary high temperatures cause the chlorides formed with the active material to sublime from the carrier.

Compounds containing lead and other heavy metals pass through as aerosols and cover up the effective surface area of the catalyst.

(iii) Performance Potential. Catalytic oxidation could economically dispose of all combustible fumes that are free of appreciable amounts of unburnable solids, which tend to foul the catalyst bed. However, the catalytic afterburners are most suited for the removal of the components that will vaporize, since because of their greater mass and lack of diffusional movement, liquid droplets are not likely to contact the catalytic surface to any appreciable degree—for the catalytic burner to operate effectively, the waste stream must not contain materials that will poison the selected catalytic system.

(3) Potential Applications

The emission of combustible pollutants, either gaseous or particulate, is common to a wide variety of chemical and manufacturing processes. The release of these materials may be constant, intermittent, or cyclical in volume, temperature, and degree of contamination. The effluents may consist entirely of compounds of one specific type, although they are usually heterogeneous. In some cases, secondary pollutants, such as sulfur dioxide, nitrogen oxides, and hydrogen chloride, may be emitted following incineration. Under such circumstances, secondary effluent treatment methods such as alkalized wet scrubbers may be desirable.

The following processes are typical sources of organic gases and particulates:

Industrial dryers	Organic chemical production
Food product ovens	Synthetic rubber manufacturing
Solid waste incineration	Asphalt processing
Coke ovens	Fat rendering
Enamel backing furnaces	Fat frying of foods
Foundry core ovens	Petroleum refining
Paper coating and impregnation equipment	Paper pulp manufacture
	Gasoline distribution
Paint, varnish and lacquer manufacture and use	Degreasing

Dry cleaning	Resin curing ovens
Coffee roasting	Electrode curing ovens
Grain milling	Sweage sludge drying
Carbon black manufacture	Refuse composting

The list is by no means inclusive, but suggests the wide variety of sources which can be considered for afterburner-type pollution control. Quantification of the characteristics of the effluent from these processes for afterburner control will be made more difficult by the highly variable concentration due to dilution in the fume collection system. The information regarding composition of these pollutant streams as they relate to afterburners would include consideration of inlet temperature, concentration, toxicity, secondary pollutant formation following combustion, photochemical reactivity, heating value, nuisance value, and the degree to which the species may inhibit or enhance the combustion or catalysis processes.

3. Incinerator Economics

As presented in Chap. 1, incineration is often the highest-cost approach to waste management. Consequently, an understanding of the economics of incineration is an important part of engineering and management decision making. Unfortunately, providing definitive guides to economic analysis is as difficult as for the counterpart in design.

a. General

It would be a gross simplification, in most cases, to indicate that incineration system capital costs could be reduced to a simple table or nomograph. Usually the designs are highly customized, reflecting unique waste handling, ducting, local regulatory requirements, degree of automatic control, enclosure aesthetics, etc., which greatly affect the final system cost, even if the basic incinerator itself is of rather predictable cost. Similarly, operating costs reflect staffing practices, localized labor and utility costs, localized unit values for byproduct materials or energy, etc.

For this reason, the sections below will emphasize the elements of cost analysis as applied to incineration rather than simplified overall numbers. The experienced engineer will recognize this as not an evasion, but rather an explicit recognition of the dangers of oversimplification.

(1) Capital Investment

For many prospective incinerator owners, the initial capital investment is the crucial issue. This is particularly true for industry, where the return on invested capital is often the prime measure of business performance. Table 48 indicates the major elements of capital expenditure. Note also that both purchased equipment and installation cost (the latter can be over 200% of the equipment cost alone) should be evaluated.

TABLE 48

Elements of Capital Cost for Incineration Systems

I. Incineration system

    A. Waste conveyance

        1. Open or compaction vehicles, commercial containers

        2. Special design containers

        3. Piping, ducting, conveyors

    B. Waste storage and handling at incinerator

        1. Waste receipt and weighing

        2. Pit and crane, floor dump and front-end loader

        3. Holding tanks, pumps, piping

    C. Incinerator

        1. Outer shell

        2. Refractory

        3. Incinerator internals (grates, catalyst)

        4. Burners

        5. Fans and ducting (forced and induced draft)

        6. Flue gas conditioning (water systems, boiler systems)

        7. Air pollution control

        8. Stacks

        9. Residue handling

      10. Automatic control and indicating instrumentation

      11. Worker sanitary, locker, and office space

II. Auxiliary systems

    A. Buildings, roadways, parking areas

    B. Special maintenance facilities

TABLE (Cont.)

    C. Steam, electrical, water fuel, and compressed air supply

    D. Secondary pollution control

        1. Residue disposal (landfill, etc.)

        2. Scrubber wastewater treatment

III. Nonequipment expenses

    A. Engineering fees

    B. Land costs

    C. Permits

    D. Interest during construction

    E. Spare parts inventory (working capital)

    F. Investments in operator training

    G. Start-up expenses

    H. Technology fees to engineers, vendors

---

(2) Operating Costs

Although capital investment is an important aspect, the actual total unit cost, allowing for the cost of capital, but including all operating costs, is a more incisive measure of economic impact. For example, contract hauling, typically, requires little or no investment by the owner, but may represent an unacceptably high unit cost for disposal. Similarly, high energy scrubbers are lower in capital cost than electrostatic precipitators but, at equivalent efficiency, consume so much power that their cost per unit of gas cleaned is much higher. Typical elements of operating cost for incineration systems are shown in Table 49.

(3) Project Comparisons

As a preferred method of analysis, a plant total operating cost and investment schedule should be developed over the projected facility life-time. Replacement investments for short-lived equipment (such as land-fill bulldozers) or major overhauls should be included. Application of present worth factors, year by year, to the net annual costs and investments, produces a "total venture cost." The total venture cost represents that amount of cash needed at the start of a project which would totally provide for the financial needs of the project. The discount rate used is that by

TABLE 49

Elements of Operating Cost for Incineration Systems

I. Fixed costs (credits)

    A. Repayment of debt capital

    B. Payment of interest on outstanding capital

    C. Tax credits for depreciation

II. Semivariable costs

    A. Labor (including supervision) with overheads

    B. Insurance

    C. Operating supplies

    D. Maintenance and maintenance supplies

III. Variable costs (credits)

    A. Steam usage (or credits)

    B. Electricity

    C. Water supply and sewerage fees

    D. Oil or natural gas fuels

    E. Compressed air

    F. Chemicals (catalysts, water treatment)

    G. Byproduct credits

    H. Disposal fees

---

which the prospective owner values capital or can obtain capital. By this means true <u>net cost</u> comparisons can be made, on an equitable basis, between high capital, low operating cost projects, and those which are low capital, high operating cost.

b. Economics for Selected Incineration Systems

(1) Municipal Incinerators

The costs for municipal incinerators vary widely. The estimates shown in Tables 50 and 51 were developed for 1975 in the Richmond, Virginia area [173]. The estimates are based on a throughput of 910 and 1360 ton/day and a design peak capacity of 1335 and 1700 ton/day, respectively. The air pollution control levels correspond to the most stringent in the United States at the time: 0.03 grains/sft^{3*} corrected to 50% excess air. Amortization was charged at 7% for all capital assets; the amortization period varies as specified, but never exceeds 20 years. Residue disposal landfills include leachage collection. Land, site development, byproduct distribution costs, and indirect costs (engineering, construction interest, working capital, taxes) are not included.

(2) Small Multiple-Chamber Incinerators

A recent volume by Corey et al. [127] has shown costs from $2300 to $31,600 for incinerators ranging in capacity from 23 to 910 kg/hr, respectively. These costs (developed in 1966 and including gas scrubbers) are particularly related to the cost of labor and only secondarily to the choice or present costs of refractory, the principal material expense. It is highly recommended, therefore, that local furnace contractors be consulted in developing a current price. Also, the costs for foundations, stacks, and utilities should be estimated and added.

TABLE 50

Capital and Operating Cost Estimates for
Conventional Incineration

BASIS

Incinerator

7 day/week, 50 week/year
three shifts per day, four shifts required per week
processing equipment availability of 80%
source: Baltimore Incinerator No. 4, Baltimore, Md.

Landfill

80% weight reduction due to incineration
residue density of 1900 kg/m^3

6350 METRIC TONS PER WEEK PLANT

Additional basis

Incinerator

three 360 ton/day (24-hr) incinerators

*Gains per standard [volume (of a gas) at standard conditions of temperature and pressure] cubic feet.

TABLE 50 (Cont.)

Landfill

total landfill site: 15.4 hectares
fill depth, 6.1 m (residue plus cover)

CAPITAL

Incinerator

three 360 ton/day incinerators[a]

incinerator	$2,650,000
stack	560,000
fans, ducts	760,000
piping	850,000
electrical, instrumentation	1,240,000
conveyors	1,290,000
Subtotal	$7,350,000
three electrostatic precipitators	3,730,000
building[b] 67,970 m^3 × $0.1075/m^3	9,050,000
three bridge cranes	1,500,000
Total incinerator capital cost (TICC)	$21,630,000

Residue landfill

site preparation	$40,000
leachate collection system	
PVC liner	207,400
10.2 cm PVC pipe	138,200
15.2 cm sand/gravel	278,500
equipment	$80,000
Total landfill capital cost	742,100
Total capital cost	$22,372,000

Operating

manpower (48 people)	$932,000
electrical power	498,000
water and sewer	134,000
heating and auxiliary fuel	13,000
maintenance (5% of TICC)	1,082,000
residue disposal	78,000
Total annual operating cost	$2,737,000

TABLE 50 (Cont.)

Owning

  amortization (7%, 20 years)

incinerator	$2,042,000
landfill	97,000
Total annual owning cost	$2,139,000
ferrous metal (net salvage revenue)[c]	($246,000)
Total net annual owning and operating cost	$4,630,000
unit capital cost	$24,700/ton daily capacity
unit cost	$14.57/ton

9530 TONS PER WEEK PLANT

Additional basis

  Incinerator

    four 410 ton/day (24-hr) incinerators

  Landfill

    total landfill site: 22 hectares
    fill depth, 6.1 m (residue plus cover)

CAPITAL

  Incinerator

    four 410 ton/day incinerators[a]

incinerator	$3,830,000
stack	810,000
fans, ducts	1,110,000
piping	1,230,000
electrical, instrumentation	1,780,000
conveyors	1,860,000
Subtotal	$10,610,000
four electrostatic precipitators	$5,370,000
building[b] 88,400 m^3 × $0.1075/m^3	11,860,000
four bridge cranes	2,000,000
Total incinerator capital cost (TICC)	$29,840,000

TABLE 50 (Cont.)

Residue landfill	
site preparation	$55,000
leachate collection system	
PVC liner	291,600
10.2 cm PVC pipe	194,400
15.2 cm sand/gravel	388,800
equipment	118,500
Total landfill capital cost	1,048,000
Total capital cost	$30,888,000
Operating	
manpower (58 people)	$1,130,000
electrical power	747,000
water and sewer	201,000
heating and auxiliary fuel	18,000
maintenance (5% of TICC)	1,490,000
residue disposal	106,000
Total annual operating cost	$3,692,000
Owning	
amortization (7%, 20 years)	
incinerator	$2,817,000
landfill	131,000
Total annual owning cost	$2,948,000
ferrous metal (net salvage revenue)[c]	($369,000)
Total net annual owning and operating cost	$6,271,000
unit capital cost	$22,700/ton daily capacity
unit cost	$13.15/ton

[a]Total installed costs including contractors, overhead, and profit.
[b]Includes foundations, pit, building enclosures, office space, weight scale, etc.
[c]Ferrous metal estimated at 7.8% of incoming refuse; 90% recovery and $11.00 per ton of scrap net revenue assumed.

TABLE 51

Capital and Operating Cost Estimates for
Heat Recovery Incineration

BASIS

Heat recovery incinerator

7 day/week, 50 week/year
three shifts per day, four shifts required per week
processing equipment availability of 80%
source: Saugus, Mass. installation

Landfill of residue

80% weight reduction due to incineration
residue density of 1900 kg/m^3

6350 METRIC TONS PER WEEK PLANT

Additional basis

Heat recovery incinerator

three 360 ton/day (24-hr) incinerators

Landfill

total landfill site: 15.4 hectares
fill depth, 6.1 m (residue plus cover)

CAPITAL

Heat recovery incinerator

refuse receiving and handling	$3,980,000
refuse combustion[a]	14,870,000
air pollution control	2,420,000
ash handling and disposal	2,170,000
utilities and miscellaneous	850,000
boiler feedwater supply and treatment	530,000
stand-by boilers	650,000
Total incinerator capital cost (TICC)	$25,470,000

Residue landfill

site preparation	$40,000
PVC liner	207,400
10.2 cm PVC pipe	138,200
15.2 cm sand/gravel	276,500

TABLE 51 (Cont.)

equipment	$80,000
Total landfill capital cost	$742,100
Total capital cost	$26,212,000

Operating

manpower	$980,000
electric power	456,000
water, sewerage	110,000
heating, auxiliary fuel	11,000
maintenance (5% TICC)	1,274,000
overhead, insurance (1% TICC)	255,000
residue disposal	78,000
Total annual operation cost	$3,164,000

Owning

amortization (7%, 20 years)

heat recovery incinerator	$2,404,000
landfill	97,000
Total annual owning cost	$2,501,000
ferrous metal (net salvage revenue)[a]	($246,000)
Total net annual owning and operating cost[b]	$5,419,000
unit capital cost	$28,860/ton daily capacity
unit cost[c]	$17.05/ton

9530 TONS PER WEEK PLANT

Additional basis

Heat recovery incinerator

four 410 ton/day (24-hr) incinerators

Landfill

total landfill site: 22 hectares
fill depth, 6.1 m (residue plus cover)

TABLE 51 (Cont.)

CAPITAL

Heat recovery incinerator

refuse receiving and handling	$5,730,000
refuse combustion[c]	21,440,000
air pollution control	3,490,000
ash handling and disposal	3,130,000
utilities and miscellaneous	1,270,000
boiler feedwater supply and treatment	790,000
stand-by boilers	750,000
Total incinerator capital cost (TICC)	$36,600,000

Residue landfill

site preparation	$55,000
PVC liner	291,600
10.2 cm PVC pipe	194,400
15.2 cm sand/gravel	388,800
equipment	$118,500
Total landfill capital cost	$1,048,300
Total capital cost	$37,648,300

Operating

manpower	$1,190,000
electric power	684,000
water, sewerage	166,000
heating, auxiliary fuel	16,000
maintenance (5% TICC)	1,830,000
overhead, insurance (1% TICC)	366,000
residue disposal	106,000
Total annual operating cost	$4,358,000

Owning

amortization (7%, 20 years)

heat recovery incinerator	$3,455,000
landfill	131,000
Total annual owning cost	$3,586,000

TABLE 51 (Cont.)

ferrous metal (net salvage revenue)	$369,000
Total net annual owning and operating cost[b]	$7,575,000
unit capital cost	$27,650/ton daily capacity
unit cost[b]	$15.90/ton

[a]Ferrous metal estimated at 7.8% of incoming refuse; 90% recovery and $11.00 per ton of scrap net revenue assumed.
[b]Revenue from steam sales are not included but would approximate $10.40 per ton (at $6.00/million kcal).
[c]Includes building, chutes, grates, furnaces, etc.

(3) Multiple Hearth Municipal Sludge Incinerators

A recent cost analysis [174] produced the cost curves shown in Fig. 44.

(4) Fluid Bed Incinerators

There is very little published data on the operation and mainteance (O/M) costs of fluid bed incineration. Albertson [175] has reported costs for fluidized systems operating mainly with primary sludge. The principal deterrent to development of field O/M costs has been the fact that most previous installations have been at smaller plants where record keeping has not been sufficient. One study collected capital cost data for several size systems [176]. In terms of February, 1968 dollars, the average capital cost of a 500 lb/hr dry solids system including the fluidized bed unit and centrifuge but excluding the preheater, buildings, installation, or engineering costs was $195,000. A 5,000 lb/hr system on the same basis had a cost of $706,000, while a 250 lb/hr system had a cost of $122,000. One comparison [177] of the costs of multiple-hearth and fluidized bed systems provides some insight into the inherent differences between the two systems. Capital costs for the two competing incineration systems are often essentially equivalent as demonstrated by results of bidding [178].

(5) Afterburners

A cost evaluation for afterburner devices requires consideration of many factors unique to the installation and locale. Often ducting costs and/ or the hoods to collect the fume constitute a large portion of the capital investment. Operating costs are strongly related to fuel costs and, for catalytic systems, to the frequency of catalyst replacement due to mechanical or chemical fouling. A recent comparison of the factors involved and the cost trade-offs is presented by Allenspach [179].

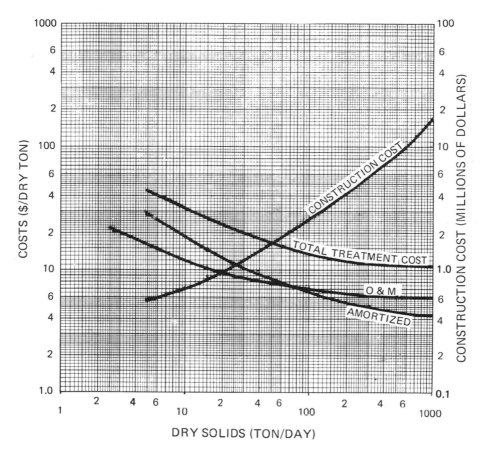

FIG. 44. Multiple hearth incineration costs: (1) Minneapolis, March, 1972. ENR Construction Cost Index of 1827; (2) amortization at 7% for 20 years; (3) labor rate of $6.25 per hour; (4) exhaust-gas scrubber and enclosing structure included; (5) costs do not include deodorization of gases: where required, add $4 to $10 per dry ton; (6) source: EPA Cost and Manpower Report and Stanley Consultants (from Ref. 174).

APPROACHES TO INCINERATOR DESIGN

How do you design an incinerator? The author wishes the answer were straightforward, the underlying principles uncluttered with contradiction and free of the need to apply judgments. This chapter, however, will only scratch the surface of the challenge of system design in an attempt to structure, if not guide, the design process.

A. CHARACTERIZE THE WASTE

Using the methods and cautions described previously, obtain the best practical characterization of the quantity and composition of the waste. Keep in mind future growth and the impact of changes in technology and economics on operational patterns and decision making.

B. LAY OUT THE SYSTEM IN BLOCKS

Too often, incineration facilities are developed in pieces with insufficient attention given to the mating of interfaces between various elements. Remember the concept of system.

C. ESTABLISH PERFORMANCE OBJECTIVES

Review present or soon to be enacted (or enforced) regulatory requirements for effluent quality. Evaluate the needs for volume reduction, residue burnout, or detoxification. Apply these to appropriate points in the facility layout.

D. DEVELOP HEAT AND
   MATERIAL BALANCES

Using the techniques developed early in this book, determine the flows
of material and energy in the waste, combustion air, and flue gases. Take
into consideration probable materials of construction, and establish reasonable limits on temperatures. Explore the impact of variations from the
"average" waste feed composition and quantity. In practice, these off-
average characteristics will generally better characterize the day to day
operating conditions.

E. DEVELOP INCINERATOR ENVELOPE

Using heat release rates per unit area and per unit volume, the overall size
of the system can be established. Use burning rate, flame length and shape,
kinetic expressions, and other analysis tools to establish the basic incinerator envelope. The final shape will depend on judgment as well as on these
calculations. Draw on the literature and the personal experience of others.
Interact with other engineers, vendors, operators, and designers of other
combustion systems with similar operating goals or physical arrangements.
Attempt to find the balance between costly overconservatism and the unfor-
tunate fact that few of the answers are tractable to definitive analysis and
computation. Particularly, talk to operators of systems. Too often the
designers speak only to one another, and the valuable insights of direct
personal experience go unheard and, worse, unasked for.

F. EVALUATE INCINERATOR DYNAMICS

Apply the jet and burner evaluation methodology, buoyancy calculations,
empirical relationships, and conventional furnace draft and pressure drop
evaluation tools to grasp, however inadequately, the dynamics of the system.

G. DEVELOP THE DESIGNS OF
   AUXILIARY EQUIPMENT

Determine the sizes and requirements of burners, fans, grates, materials
handling systems, pumps, air compressors, air pollution control systems,
and the many other auxiliary equipments comprising the system. Here
again, the caution is to be generous, protective, rugged. The cardinal
rule is to prepare for when "it" happens, not to argue about if "it" will
happen.

I. BUILD AND OPERATE

In many cases, fortunately, nature is kind: reasonable engineering designs
will function, though perhaps not to expectations. At times, plants built
using the most detailed analysis and care result in failure. Such is the lot
of workers in the complex but fascinating field of combustion and incineration.

SYMBOLS, CONVERSION FACTORS,
PERIODIC TABLE OF THE ELEMENTS

TABLE OF SYMBOLS

Symbol	Definition	Unit
A	The "frequency factor" or pre-exponential constant in the Arrhenius kinetics expression	varies
A_e	Cross-sectional area of body of cyclone combustion chamber	m^2
A_t	Tangential inlet area of cyclone combustion chamber	m^2
a	Constant a, a_1, a_2, etc.	—
B	Mass transfer driving force	—
C_x	Drag coefficient	—
c	Concentration: c_0, initial or at nozzle; $\overline{c}_m$, mean at jet centerline; $\overline{c}$, on radius of jet	mol/m^3
c_p	Specific heat: c_p°, at zero pressure; $c_{p,av}^\circ$, average at zero pressure between given limits of temperature	$kcal\ kg^{-1}\ {}^\circ C^{-1}$
D	Combustion chamber diameter	m
D_s	Stack diameter	m

TABLE OF SYMBOLS (Cont.)

Symbol	Definition	Unit
d	Diameter: d_0, nozzle or initial; d_c, combustion chamber; d_e, exit throat; d_0', effective nozzle diameter	m
E	Energy (enthalpy) change on reaction	cal/mol
ΔE	Activation energy (enthalpy)	cal/mol
F	Force: F_B, buoyant force; F_D, drag force	newtons
F_e	Expansion draft loss in a stack	atm
F_s	Friction draft loss in a stack	atm
f_a	Mole fraction of component a in a mixture	—
f_s	Mole fraction of nozzle gas in a stoichiometric mixture	—
$\dot{G}_0$	Nozzle flux of linear momentum in a swirling jet	$kg\ m^2\ sec^{-2}$
g	Acceleration of gravity	m/sec^2
H	Enthalpy (total)	kcal
ΔH	Heat (enthalpy) of reaction	kcal/kg mol
ΔH_c	Heat (enthalpy) of combustion	kcal/kg
ΔH_v	Heat (enthalpy) of vaporization	kcal/kg
H_s	Enthalpy at reference temperature (here, the temperature of the liquid at the surface of a burning pool	kcal/kg
h	Enthalpy per mole	kcal/kg mol
Δh	Sensible heat	kcal
h_c	Heat transfer coefficient by convection	$kcal\ hr^{-1}\ m^{-2}\ °C^{-1}$
HHV	Higher heating value (water condensed)	kcal/kg mol
J	Entrainment coefficient of a jet ($\bar{v}_m/\bar{u}_m$)	—
K	Time constant for chemical reaction	sec^{-1}
K_d	Diffusional rate constant (soot oxidation)	$g\ cm^{-2}\ sec^{-1}$
K_p	Equilibrium constant (pressure)	varies

TABLE OF SYMBOLS (Cont.)

Symbol	Definition	Unit
K_s	Kinetic rate constant (soot oxidation)	$g\ cm^{-2}\ sec^{-1}$
k	Constant	varies
L	Flame length	m
$\dot{L}_0$	Nozzle flux of angular momentum in a swirling jet	$kg\ m^2\ sec^{-2}$
L_j	Jet penetration distance	m
L_T	Furnace length	m
L_s	Height of a stack above the breeching	m
ℓ	Distance along curved path of a deflected jet	m
LHV	Lower heating value (water uncondensed)	kcal/kg mol
M	Momentum ratio for jet in crossflow	—
$M^\circ_{c_{P,av}}$	Average molar heat capacity at zero pressure between given limits of temperature	$kcal\ kg\ mol^{-1}\ {}^\circ C^{-1}$
M_0	Molecular weight of nozzle gas	kg/kg mol
M_a	Molecular weight of air	kg/kg mol
M_r	Recirculating mass flow relative to nozzle flow	—
m	Surface density	g/cm^2
$\dot{m}$	Mass flow rate: $(\dot{m}_x)$, at a distance x in a jet; $(\dot{m}_0)$, at the nozzle; $(\dot{m}_r)$, recirculating	kg/sec
$\dot{m}''$	Mass burning rate per unit area	$g\ cm^{-2}\ sec^{-1}$
MW	Molecular weight	kg/mol
N	Number of nozzles in a linear array	—
N_f	Nitrogen in fuel	mol
N_{Re}	Reynolds number $(du\rho/\mu)$	—
N_s	Swirl number for axial swirl burners	—
N_{sc}	Swirl number for cyclone chambers	—

TABLE OF SYMBOLS (Cont.)

Symbol	Definition	Unit
n	Number of kilogram moles	kg mol
P	Pressure: (ΔP), differential; (ΔP_i), across inlet; (ΔP_{cc}), across combustion chamber; (P_{ND}), natural draft; (P_0), barometric (ambient) pressure	kg m^{-2}, atm
p	Partial pressure	atm
Q_T	Volume flow of overfire air	m^3/min
$\dot{Q}_V$	Heat release rate per unit volume	kcal m^{-3} sec^{-1}
q	Rate of consumption per unit geometric external surface area	g cm^{-2} sec^{-1}
R, R'	Universal gas constant	varies (see Table 1)
r	Radius: (r_d), droplet radius	m
r, r'	Chemical reaction rate	varies
S	Slope of a kiln	m/m
s	Spacing between jet centers in a linear array	m
T	Temperature: (T_0), initial or at nozzle; (T_a), ambient; (T_F), adiabatic flame; (T_R), reference; (T_M), mixture; (T_d), droplet	°C, K
t	Time: residence time; (t_b), particle burnout time	sec
u	Velocity: (u_0), at nozzle or initial; $(\bar{u}_m)$, time averaged mean on jet centerline; $(\bar{u}_0)$, time averaged mean at nozzle; (u'), r.m.s. fluctuating component in axial direction; $(\bar{u}_1)$, mean crossflow; $(\bar{u}_e)$, average at exit; (u_f), final; (u_t), terminal	m/sec
V	Volume	m^3
$\dot{V}$	Volume flow rate	m^3/sec
v	Velocity: (v'), r.m.s. fluctuating component in radial velocity in a jet; $(\bar{v}_m)$, mean radial velocity into jet	m/sec

TABLE OF SYMBOLS (Cont.)

Symbol	Definition	Unit
W	Radiative emissive power: (W_B), of a black body	$kcal\ m^{-2}\ hr^{-1}$
w	Velocity tangent to a circle centered on the axis of a swirling jet; (w'), r.m.s. fluctuating component in tangential direction; (w_{in}), tangential inlet velocity to cyclone combustor	m/sec
x	A variable, defined as used but often a distance in the x direction; (x^+), distance to end of zone of establishment in a jet; (x_p), impingement point of a ducted jet	varies
Y	Conversion ratio for fuel nitrogen into NO	—
y	A variable, defined as used but often a distance in the y direction; (y^+), distance to end of a zone of establishment in a jet; (y_0), width of a slot jet	varies
Z	Length of a linear jet array	m

Greek Symbols

Symbol	Definition	Unit
α	Thermal diffusivity	m^2/sec
α_0	Jet deflection angle (initial)	degrees or radians
β	Burnedness, the relative degree of completeness of combustion	—
γ	Exponential function $(1/\zeta)\ exp\ (-\zeta)$; (γ_1), exponential integral	—
Δ	A prefix denoting change or difference in a quantity	varies
δ	Angle of attack between jet and crossflow vector	degrees or radians
ϵ	Emissivity for radiative heat transfer	—
ζ	Nondimensional temperature $\Delta E/RT$	—
θ	Kilograms of oxidant per kilogram of fuel for a stoichiometric mixture	—

TABLE OF SYMBOLS (Cont.)

Symbol	Definition	Unit
θ_m	Moles of reactant per mole of product in a stoichiometric mixture	—
λ	Thermal conductivity	$\text{kcal m}^{-2}\text{sec}^{-1}(°C/m)^{-1}$
μ	Viscosity; (μ_0), initial	$\text{g cm}^{-1}\text{ sec}^{-1}$, $\text{kg m}^{-1}\text{ sec}^{-1}$
ν	Kinematic viscosity	cm^2/sec
ξ	Pressure loss coefficient	—
π	Constant (3.14159...)	—
ρ	Density: (ρ_0), initial or at nozzle; (ρ_a), ambient; (ρ_e), liquid; (ρ_f), at flame temperature; (ρ_{iso}), isothermal case; (ρ_s), of particles; (ρ_c), of char; (ρ_v), of virgin material	kg/m^3
σ	Stephan-Boltzmann constant	$\text{kcal m}^{-2}\text{ hr}^{-1}\text{ (K)}^{-4}$
Φ	Mixture ratio (mass air/mass fuel) $= 1/\theta$	—
ϕ	Jet half-angle; (ϕ_0), without swirl; (ϕ_s), with swirl	radians
Ω	Dilution ratio of a jet $(\overline{c}/\overline{c}_0)$	—
ω	Parameter in recirculating jet flow	—

TABLE OF CONVERSION FACTORS

Quantity	Unit A	×	Factor	=	Unit B
Length	angstrom		1×10^{-10}		meter
	foot		0.3048		meter
	foot		0.3333		yard
	inch		2.54		centimeter
	meter		100		centimeter
	meter		3.2808		foot
	meter		1.0936		yard
	micrometer		1×10^{-6}		meter
Area	acre		4046.9		square meter
	square centimeter		0.155		square inch
	square inch		6.4516		square centimeter
	square foot		929.03		square centimeter
	square foot		929.03		square centimeter
	square foot		0.0929		square meter
	square meter		10.7639		square foot
	square yard		9		square foot
Volume	barrel (oil)		42		gallon
	barrel (oil)		159.983		liter
	cubic foot		2.8317×10^4		cubic centimeter
	cubic foot		7.481		gallon
	cubic foot		0.028317		cubic meter
	cubic meter		1×10^6		cubic centimeter
	cubic yard		0.7646		cubic meter
	gallon (U.S.)		0.8327		gallon (British)
	gallon		3.7853		liter
	gallon (water)		8.337		pound
	liter		0.03532		cubic foot
	liter		0.001		cubic meter
	liter		0.26418		gallon
	liter		1.0567		quart
Mass	grain		0.0648		gram
	grain		1.4286×10^{-4}		pound
	kilogram		1000		gram
	kilogram		2.2046		pound
	kilogram		0.001		tonne
	microgram		1×10^{-6}		gram
	pound		453.59		gram
	pound		0.4536		kilogram
	tonne		1000		kilogram
	tonne		2204.62		pound

TABLE OF CONVERSION FACTORS (Cont.)

Quantity	Unit A	×	Factor	=	Unit B
	ton (short)		2000		pound
	ton (long)		2240		pound
Time	day		1440		minute
	day		8.64×10^4		second
	minute		60		second
	year		8760		hour
Velocity	centimeter/ second		1.9685		foot/minute
	foot/minute		0.00508		meter/second
	foot/second		0.3048		meter/second
	meter/minute		0.0547		foot/second
	meter/second		3.281		foot/second
	mile/hour		0.447		meter/second
Flow	cubic centimeters/ second		0.002119		cubic feet/minute
	cubic feet/minute		0.1247		gallon/second
	cubic feet/minute		0.4720		liter/second
	cubic feet/second		4.72×10^{-4}		cubic meter/second
	cubic meter/ second		35.31		cubic feet/second
	cubic meter/ second		2118.6		cubic feet/minute
	cubic meter/ minute		16.667		liter/second
	gallon/minute		0.06308		liter/second
Pressure	atmosphere		14.696		pound/square inch
	atmosphere		29.921		inch of mercury
	atmosphere		760		millimeter of mercury
	atmosphere		1033.2		gram/square centimeter
	atmosphere		33.899		foot of water
	inch of water		1.8683		millimeter of mercury
	inch of water		0.002458		atmosphere
	pound/square inch		0.06805		atmosphere
Energy	Btu		0.25198		kilocalorie
	joule		2.389×10^{-4}		kilocalorie
	kilocalorie		3.9865		Btu
	kilocalorie		1000		calorie
	kilocalorie		0.0015593		horsepower hour
	kilocalorie		0.001628		kilowatt hour

TABLE OF CONVERSION FACTORS (Cont.)

Quantity	Unit A	$\times$	Factor	$=$	Unit B
Temperature	$\frac{5}{9}$ ($^\circ$F $-$ 32)				$^\circ$C
	$\frac{9}{5}$ $^\circ$C + 32				$^\circ$F
	$^\circ$C + 273.18				$^\circ$K
	$^\circ$F + 460				$^\circ$R
Heat capacity	Btu/pound $^\circ$F		1.0		calorie/gram $^\circ$C
Heat content	Btu/pound		0.5556		calorie/gram
	Btu/pound		0.5556		kilocalorie/kilogram
	kilocalorie/ kilogram		1.8		Btu/pound
	joule/gram		0.4301		Btu/pound
Thermal conductivity	Btu/second square foot $^\circ$F/ inch		1.2404		calorie/second square centimeter $^\circ$C/ centimeter
	Btu/second square foot $^\circ$F/ foot		1.487		kilocalorie/nour square meter $^\circ$C/meter
Viscosity	gram (weight) second/square centimeter		980.665		poise
	poise		0.01		centipoise
Miscellaneous Particulate loading in a gas	grain/cubic foot		2.29×10^6		microgram/cubic meter
Stephen– Boltzmann constant	0.173×10^{-8} Btu/ hour square foot $^\circ$F^4				4.92×10^{-8} kilocalorie/ hour square meter $^\circ$K^4
Gravitation– al constant	32.2 feet/second/ second				980.7 centimeters/ second/second

TABLE OF INTERNATIONAL ATOMIC WEIGHTS

Substance	Symbol	Atomic weight	Atomic number	Substance	Symbol	Atomic weight	Atomic number
Actinium	Ac	227	89	Neodymium	Nd	144.27	60
Aluminum	Al	26.97	13	Neon	Ne	20.183	10
Americium	Am	241	95	Neptunium	Np	237	93
Antimony	Sb	121.76	51	Nickel	Ni	58.69	28
Argon	A	39.944	18	Niobium	Nb	92.91	41
Arsenic	As	74.91	33	Nitrogen	N	14.008	7
Astatine	At	211	85	Osmium	Os	190.2	76
Barium	Ba	137.36	56	Oxygen	O	16.0000	8
Beryllium	Be	9.02	4	Palladium	Pd	106.7	46
Bismuth	Bi	209.00	83	Phosphorus	P	30.98	15
Boron	B	10.82	5	Platinum	Pt	195.23	78
Bromine	Br	79.916	35	Plutonium	Pu	239	94
Cadmium	Cd	112.41	48	Polonium	Po	210	84
Calcium	Ca	40.08	20	Potassium	K	39.096	19
Carbon	C	12.010	6	Praseodymium	Pr	140.92	59
Cerium	Ce	140.13	58	Promethium	Pm	147	61
Cesium	Cs	132.91	55	Protactinium	Pa	231	91
Chlorine	Cl	35.457	17	Radium	Ra	226.05	88
Chromium	Cr	52.01	24	Radon	Rn	222	86
Cobalt	Co	58.94	27	Rhenium	Re	186.31	75
Copper	Cu	63.54	29	Rhodium	Rh	102.91	45
Curium	Cm	242	96	Rubidium	Rb	85.48	37

Name	Symbol	Atomic Weight	Atomic Number
Dysprosium	Dy	162.46	66
Erbium	Er	167.2	68
Europium	Eu	152.0	63
Fluorine	F	19.00	9
Francium	Fr	223	87
Gadolinium	Gd	156.9	64
Gallium	Ga	69.72	31
Germanium	Ge	72.60	32
Gold	Au	197.2	79
Hafnium	Hf	178.6	72
Helium	He	4.003	2
Holmium	Ho	164.94	67
Hydrogen	H	1.0080	1
Indium	In	114.76	49
Iodine	I	126.92	53
Iridium	Ir	193.1	77
Iron	Fe	55.85	26
Krypton	Kr	83.7	36
Lanthanum	La	138.92	57
Lead	Pb	207.21	82
Lithium	Li	6.940	3
Lutetium	Lu	174.99	71
Magnesium	Mg	24.32	12
Manganese	Mn	54.93	25
Mercury	Hg	200.61	80
Molybdenum	Mo	95.95	42
Ruthenium	Ru	101.7	44
Samarium	Sm	150.43	62
Scandium	Sc	45.10	21
Selenium	Se	78.96	34
Silicon	Si	28.06	14
Silver	Ag	107.880	47
Sodium	Na	22.997	11
Strontium	Sr	87.63	38
Sulfur	S	32.066	16
Tantalum	Ta	180.88	73
Technetium	Tc	99	43
Tellurium	Te	127.61	52
Terbium	Tb	159.2	65
Thallium	Tl	204.39	81
Thorium	Th	232.12	90
Thulium	Tm	169.4	69
Tin	Sn	118.70	50
Titanium	Ti	47.90	22
Tungsten	W	183.92	74
Uranium	U	238.07	92
Vanadium	V	50.95	23
Xenon	Xe	131.3	54
Ytterbium	Yb	173.04	70
Yttrium	Y	88.92	39
Zinc	Zn	65.38	30
Zirconium	Zr	91.22	40

COMBUSTION PROPERTIES OF COAL, OIL, NATURAL GAS, AND OTHER MATERIALS

Often in waste incineration, conventional fossil fuels are burned. The data in Tables B-1 to B-7 will then be useful to carry out the relevant energy and material balances.

Heat of Combustion Data (Table B-8) are presented for a variety of chemical substances and materials in common use in industry and/or commonly appearing in waste streams. Note that the values presented are for the lower heating value (water as a gas). See also the several tables in the text (especially Tables 34, 36 and 38).

Inflammability limits and ignition temperatures in Table B-9 are useful in evaluating the hazards in transporting or storing materials.

TABLE B-1

Classification of Coals by Rank[a] (ASTM D 388)

Class	Group	Fixed carbon limits, % (dry, mineral-matter-free basis)		Volatile matter limits, % (dry, mineral-matter-free basis)		Calorific value limits, kcal/kg (moist,[b] mineral-matter-free basis)		Agglomerating character
		Equal or greater than	Less than	Greater than	Equal or less than	Equal or greater than	Less than	
I. Anthracite	1. Meta-anthracite	98	—	—	2	—	—	Nonagglomerating
	2. Anthracite	92	98	2	8	—	—	
	3. Semianthracite[c]	86	92	8	14	—	—	
II. Bituminous	1. Low volatile bituminous coal	78	86	14	22	—	—	
	2. Medium volatile bituminous coal	69	78	22	31	—	—	
	3. High volatile A bituminous coal	—	69	31	—	7778[d]	—	Commonly agglomerating[e]
	4. High volatile B bituminous coal	—	—	—	—	7222[d]	7778	
	5. High volatile C bituminous coal	—	—	—	—	6389 / 5833[e]	7222 / 5833	Agglomerating

						Calorific value limits, mineral-matter-free kcal/kg	
	1. Subbituminous A coal	—	—	—	—	5833	6389
III. Subbituminous	2. Subbituminous B coal	—	—	—	—	5278	5833
	3. Subbituminous C coal	—	—	—	—	4611	5278
	1. Lignite A	—	—	—	—	3500	4611
IV. Lignite	2. Lignite B	—	—	—	—	—	3500

Nonagglomerating

a This classification does not include a few coals, principally nonbanded varieties, which have unusual physical and chemical properties and which come within the limits of fixed carbon or calorific value of the high-volatile bituminous and subbituminous ranks. All of these coals either contain less than 48% dry, mineral-matter-free fixed carbon or have more than 8610 moist, mineral-matter-free kcal/kg (higher heating value).

b Moist refers to coal containing its natural inherent moisture but not including visible water on the surface of the coal.

c If agglomerating, classify in low-volatile group of the bituminous class.

d Coals having 69% or more fixed carbon on the dry, mineral-matter-free basis shall be classified according to fixed carbon, regardless of calorific value.

e It is recognized that there may be nonagglomerating varieties in these groups of the bituminous class, and there are notable exceptions in high volatile C bituminous group.

TABLE B-2

Seventeen Selected U.S. Coals Arranged in Order of ASTM Classification[a]

No.	Coal rank			County	State	Coal analysis, bed moisture basis					
	Class	Group				M	VM	FC	A	S	kcal
1	I	1	Pa.	Schuylkill	Pa.	4.5	1.7	84.1	9.7	0.77	7081
2	I	2	Pa.	Lackawanna	Pa.	2.5	6.2	79.4	11.9	0.60	7181
3	I	3	Va.	Montgomery	Va.	2.0	10.6	67.2	20.2	0.62	6625
4	II	1	W. Va.	McDowell	W. Va.	1.0	16.6	77.3	5.1	0.74	8175
5	II	1	Pa.	Cambria	Pa.	1.3	17.5	70.9	10.3	1.68	7667
6	II	2	Pa.	Somerset	Pa.	1.5	20.8	67.5	10.2	1.68	7622
7	II	2	Pa.	Indiana	Pa.	1.5	23.4	64.9	10.2	2.20	7667
8	II	3	Pa.	Westmoreland	Pa.	1.5	30.7	56.6	11.2	1.82	7403

					M	VM	FC	A	S	kcal
9	II	3	Ky.	Pike	2.5	36.7	57.5	3.3	0.70	8044
10	II	3	Ohio	Belmont	3.6	40.0	47.3	9.1	4.00	7139
11	II	4	Ill.	Williamson	5.8	36.2	46.3	11.7	2.70	6617
12	II	4	Utah	Emery	5.2	38.2	50.2	6.4	0.90	7000
13	II	5	Ill.	Vermilicn	12.2	38.8	40.0	9.0	3.20	6300
14	III	1	Mont.	Musselshell	14.1	32.2	46.7	7.0	0.43	6189
15	III	2	Wyo.	Sheridan	25.0	30.5	40.8	3.7	0.30	5192
16	III	3	Wyo.	Campbell	31.0	31.4	32.8	4.8	0.55	4622
17	IV	1	N.D.	Mercer	37.0	26.6	32.3	4.2	0.40	4130

[a]For definition of rank classification according to ASTM requirements, see Table B-1. Data on coal (bed moisture basis): M = equilibrium moisture, %; VM = volatile matter, %; FC = fixed carbon, %; A = ash, %; S = sulfur, %; kcal = kcal/kg, high heating value.

TABLE B-3

Coal Analysis, Dry, Ash-free Bais (%)

Coals, Table B-2 No.	Volatile matter	Fixed carbon	C	H_2	O_2	N_2	kcal/kg[a]
1	2.0	98.0	93.9	2.1	2.3	0.3	8250
2	7.3	92.7	93.5	2.6	2.3	0.9	8389
3	13.6	86.4	90.7	4.2	3.3	1.0	1514
4	17.7	82.3	90.4	4.8	2.7	1.3	8706
5	19.8	80.2	89.4	4.8	2.4	1.5	8675
6	23.5	76.5	88.6	4.8	3.1	1.6	8633
7	26.5	73.5	87.6	5.2	3.3	1.4	8683

8	35.2	64.8	85.0	5.4	5.8	1.7	8481
9	39.0	61.0	85.5	5.5	6.7	1.6	8539
10	45.8	54.2	80.9	5.7	7.4	1.4	8183
11	43.8	56.2	80.5	5.5	9.1	1.6	8017
12	43.2	56.8	79.8	5.6	11.8	1.7	7922
13	49.3	50.7	79.2	5.7	9.5	1.5	8000
14	40.8	59.2	80.9	5.1	12.2	1.3	7839
15	42.8	57.2	75.9	5.1	17.0	1.6	7278
16	49.0	51.0	74.0	5.6	18.6	0.9	7206
17	45.3	54.7	72.7	4.9	20.8	0.9	6850

[a]Higher heating value.

TABLE B-4

ASTM Standard Specifications for Fuel Oils[a]

No. 1: A distillate oil intended for vaporizing pot-type burners and other burners requiring this grade of fuel.

No. 2: A distillate oil for general purpose domestic heating for use in burners not requiring No. 1 fuel oil.

No. 4: Preheating not usually required for handling or burning.

Grade of fuel oil[b]	Flash Point, °F (°C) Min	Pour Point,[e] °F (°C) Max	Water and sediment (% by volume) Max	Carbon residue on 10% bottoms, % Max	Ash (% by weight) Max	Distillation temperatures, °F (°C) 10% Point Max	90% Point Min	90% Point Max
No. 1	100 or legal (38)	0	trace	0.15	—	420 (215)	—	550 (288)
No. 2	100 or legal (38)	20[c] (-7)	0.10	0.35	—	d	540[c] (282)	640 (338)
No. 4	130 or legal (55)	20 (-7)	0.50	—	0.10	—	—	—
No. 5 (Light)	130 or legal (55)	—	1.00	—	0.10	—	—	—
No. 5 (Heavy)	130 or legal (55)	—	1.00	—	0.10	—	—	—
No. 6	150 (65)	—	2.00[g]	—	—	—	—	—

[a] Source, ASTM D 396. Recognizing the necessity for low-sulfur fuel oils used in connection with heat treatment, nonferrous metal, glass, and ceramic furnaces, and other special uses, a sulfur requirement may be specified in accordance with the following: No. 1 Grade: 0.5% max sulfur; No. 2 Grade: 0.7%; No. 4 Grade: no limit; No. 5 Grade: no limit; No. 6 Grade: no limit. Other sulfur limits may be specified only by mutual agreement between the purchaser and the seller.

[b] It is the intent of these classifications that failure to meet any requirement of a given grade does not automatically place an oil in the next lower grade unless in fact it meets all requirements of the lower grade.

[c] Lower or higher pour points may be specified whenever required by conditions of storage or use.

No. 5: (Light). Preheating may be required depending on climate and
    equipment.
No. 5: (Heavy). Preheating may be required for burning and, in cold cli-
    mates, may be required for handling.
No. 6: Preheating required for burning and handling.

| Saybolt Viscosity (sec) | | | | Kinematic Viscosity (centistokes) | | | | Gravity (deg API) | Copper strip corrosion |
| Universal at 100°F (38°C) | | Furol at 122°F (50°C) | | At 100°F (38°C) | | At 122°F (50°C) | | | |
Min	Max	Min	Max	Min	Max	Min	Max	Min	Max
—	—	—	—	1.4	2.2	—	—	35	No. 3
(32.6)[f]	(37.93)	—	—	2.0[c]	3.6	—	—	30	—
45	125	—	—	(5.8)	(26.4)	—	—	—	—
150	300	—	—	(32)	(65)	—	—	—	—
350	750	(23)	(40)	(75)	(162)	(42)	(81)	—	—
(900)	(9000)	45	300	—	—	(92)	(638)	—	—

[d]The 10% distillation temperature point may be specified at 440°F (226°C)
maximum for use in other than atomizing burners.
[e]When pour point less than 0°F is specified, the minimum viscosity shall
be 1.8 cs (32.0 sec, Saybolt Universal) and the minimum 90% point shall
be waived.
[f]Viscosity values in parentheses are for information only and not necessarily limiting.
[g]The amount of water by distillation plus the sediment by extraction shall
not exceed 2.00%. The amount of sediment by extraction shall not exceed
0.50%. A deduction in quantity shall be made for all water and sediment
in excess of 1.0%.

TABLE B-5

Selected Analyses of Fuel Oils

Grade of fuel oil	No. 1	No. 2	No. 4	No. 5	No. 6
Weight, percent					
Sulfur	0.1	0.3	0.8	1.0	2.3
Hydrogen	13.8	12.5	—	—	9.7
Carbon	86.1	87.2	—	—	85.6
Nitrogen	—	0.02	—	—	—
Oxygen	Nil	Nil	—	—	2.0
Ash	Nil	Nil	0.03	0.03	0.12
Gravity[a]					
Deg API	42	32	20	19	—
Specific gravity	0.815	0.865	0.934	0.940	—

Pour point					
°F	-35	-5	+20	+30	—
°C	-37.2	-20.5	-6.7	-1.1	—
Viscosity					
Centistokes @ 100LF (38°C)	1.8	2.4	27.5	130	—
SUS @ 100°F (38°C)	—	34	130	—	—
SSF @ 122°F (50°C)	—	—	—	30	—
Water, sediment, vol %	Nil	Nil	0.2	0.3	0.74
Heating value, kcal/kg (higher heating value)	11,006	10,794	10,478	10,422	10,167[b]

[a] American Petroleum Institute (API) gravity refers to a hydrometer scale with the following relation to specific gravity: degrees API = {141.5/[sp gr @ (15.5/15.5°C)]} - 131.5 where sp gr @ 15.5/15.5°C represents the ratio of oil density at 15.5°C (60°F) to that of water at 15.5°C.

[b] Bomb calorimeter determination.

TABLE B-6

Range of Analyses of Fuel Oils

Grade of fuel oil	No. 1	No. 2	No. 4	No. 5	No. 6
Weight, percent					
Sulfur	0.01-0.5	0.05-1.0	0.2-2.0	0.5-3.0	0.7-3.5
Hydrogen	13.3-14.1	11.8-13.9	(10.6-13.0)[a]	(10.5-12.0)[a]	(9.5-12.0)[a]
Carbon	85.9-86.7	86.1-88.2	(86.5-89.2)[a]	(86.5-89.2)[a]	(86.5-90.2)[a]
Nitrogen	Nil-0.1	Nil-0.1	—	—	—
Oxygen	—	—	—	—	—
Ash	—	—	0-0.1	0-0.1	0.01-0.5
Gravity					
Deg API	40-44	28-40	15-30	14-22	7-22
Specific gravity	0.825-0.806	0.887-0.825	0.966-0.876	0.972-0.922	1.022-0.922

Pour point					
°F	0 to −50	0 to −40	−10 to +50	−10 to +80	+15 to +85
°C	−18 to −46	−18 to −40	−28 to +10	−28 to +27	−9 to +29
Viscosity					
Centistokes @ 38°C (100°F)	1.4–2.2	1.9–3.0	10.5–65	65–200	260–750
SUS @ 38°C (100°F)	—	32–38	60–300	—	—
SSF @ 50°C (112°F)	—	—	—	20–40	45–300
Water, sediment, vol %	—	0–0.1	tr to 1.0	0.05–1.0	0.05–2.0
Heating value, kcal/kgm (higher heating value) (calculated)	10,928–11,033	10,650–10,972	10,156–10,778	10,056–10,567	9,672–10,550

aEstimated.

TABLE B-7

Selected Samples of Natural Gas from United States Fields

Analyses		Source of gas				
Constituents, vol %		Pa.	So. Cal.	Ohio	La.	Okla.
H_2	Hydrogen	—	—	1.82	—	—
CH_4	Methane	83.40	84.00	93.33	90.00	84.10
C_2H_4	Ethylene	—	—	0.25	—	—
C_2H_6	Ethane	15.80	14.80	—	5.00	6.70
CO	Carbon monoxide	—	—	0.45	—	—
CO_2	Carbon dioxide	—	0.70	0.22	—	0.80
N_2	Nitrogen	0.80	0.50	3.40	5.00	8.40

O₂ Oxygen	—	—	0.35	—	—
H₂S Hydrogen sulfide	—	—	0.18	—	—
Ultimate, wt %					
S Sulfur	—	—	0.34	—	—
H₂ Hydrogen	20.85	22.68	23.20	23.30	23.53
C Carbon	64.84	69.26	69.12	74.72	75.25
N₂ Nitrogen	12.90	8.06	5.76	0.76	1.22
O₂ Oxygen	1.41	—	1.58	1.22	—
Specific gravity (rel to air)	0.630	0.600	0.567	0.636	0.636
Higher heat value					
kcal/m³ @ 15.6°C, 1 atm	8666	8915	8577	9930	10,045
kcal/kg of fuel	11,200	12,124	12,260	12,724	12,872

TABLE B-8
Heat of Combustion for Selected Materials

I. Organic Compounds[a]

Name	Formula	Physical state	Lower heat of combustion[b]
Acetaldehyde	CH_3CHO	Liquid	279.0
Acetic acid	CH_3CO_2H	Liquid	209.4
Acetone	$(CH_3)_2CO$	Liquid	426.8
Acetylene	$CHCH$	Gas	312.0
Aniline	$C_6H_5NH_2$	Liquid	811.7
Benzaldehyde	C_6H_5CHO	Liquid	841.3
Benzene	C_6H_6	Liquid	782.3
N–Butyl alcohol	C_4H_9OH	Liquid	638.6
Carbon disulfide	CS_2	Liquid	246.6
Carbon tetrachloride	CCl_4	Liquid	37.3
Chloroform	$CHCl_3$	Liquid	89.2
Cresol (av. o, m, p)	$CH_3C_6H_4OH$	Liquid	881.9
Ethane	C_2H_6	Gas	368.4
Ethyl acetate	$CH_3CO_2C_2H_5$	Liquid	536.9
Ethyl alcohol	C_2H_5OH	Liquid	327.6
Ethylene (ethene)	CH_2CH_2	Gas	331.6
Glycerol (glycerin)	$(CH_2OH)_2$	Liquid	397.0

TABLE B-8 (Cont.)

Name	Formula	Physical state	Lower heat of combustion[b]
N-Heptane	C_7H_{16}	Liquid	1149.9
N-Hexane	C_6H_{14}	Liquid	989.8
iso-Butyl alcohol	$(CH_3)_2CH_2CH_2OH$	Liquid	638.2
iso-Propyl alcohol	$(CH_3)_2CHOH$	Liquid	474.8
Methane	CH_4	Gas	210.8
Methyl alcohol	CH_3OH	Liquid	170.9
Methylene chloride	CH_2Cl_2	Gas	106.8
Methyl ethyl ketone	$CH_3COC_2H_5$	Liquid	582.3
N-Octane	C_8H_{18}	Liquid	1302.7
Phenol	C_6H_5OH	Solid	732.2
Phthalic acid	$C_6H_4(CO_2H)_2$	Solid	771.0
Propane	C_3H_8	Gas	526.3
N-Propyl alcohol	C_3H_7OH	Liquid	480.5
Propylene	CH_3CHCH_2	Gas	490.2
Starch	$(C_6H_{10}O_5)_x$ per kg	Solid	4178.8
Toluene	$CH_3C_6H_5$	Liquid	934.2
Urea	$(NH_2)_2CO$	Solid	151.6
Xylene (av. o, m, p)	$(CH_3)_2C_6H_4$	Liquid	1089.7

TABLE B-8 (Cont.)

II. Various Substances

Name	Lower heat of combustion[c]
Asphalt	9.53
Bagasse (12% H_2O)	4.05
Bamboo (10% H_2O)	4.11
Charcoal (4% H_2O)	7.26
Coal (see Tables B-1 to B-3)	
Fats (animal)	9.50
Gases (see Tables B-7 and B-8-I)	
Iron (to Fe_2O_3)	1.58
Magnesium (to MgO)	4.73
Oil, fuel (see Tables B-4 to B-6)	
Oil, cotton seed	9.50
Paraffin	10.34
Pitch	8.40
Sulfur (rhombic)	2.20
Wood, oak (13% H_2O)	3.99
Wood, pine (12% H_2O)	4.42

[a]Source: Kharasch, U.S. Bureau of Standards, Journal of Research, 2:359 (1929).

[b]Kilocalories per gram molecular weight at 1 atm, 20°C; carbon dioxide, water vapor, and nitrogen (if present) as gases.

[c]Kilocalories per gram at 1 atm, 20°C; carbon dioxide, water vapor, and nitrogen (if present) as gases.

TABLE B-9

Inflammability Limits and Ignition Temperatures of Various Substances

Substance	State[b]	Inflammability limit (vol %)[a]		Ignition temperature (°C)	
		Lower	Upper	In air	In O$_2$
Acetic acid	g	5.4	—	—	
Acetone	g	2.55	12.80	561	—
Acetylene	g	2.50	8.00	305	—
Ammonia	g	15.50	27.00	—	—
Benzene	g	1.40	7.10	740	662
Butane	g	1.86	8.41	430	—
Butylene	g	1.65	9.95	443	—
Carbon, fixed (anthracite coal)	s	—	—	448–601	—
Carbon, fixed (bituminous coal)	s	—	—	407	—
Carbon, fixed (semibituminous coal)	s	—	—	465	—
Carbon disulfide	g	1.25	50.00	120	—
Carbon monoxide	g	12.5	71.2	644–658	637–658
Charcoal	s	—	—	650	—
Ethane	g	3.0	12.5	520–630	520–630
Ethyl alcohol	g	3.28	18.95	392	—
Ethylene	g	2.75	28.60	542–547	500–519
Ethylene oxide	g	3.00	80.00	429	—
Gasoline	l	—	—	260–426	
Heptane (n–)	g	1.1	6.7	247	—
Hexane (n–)	g	1.18	7.4	260	—
Hydrogen	g	4.	74.20	580–590	580–590
Hydrogen sulfide	g	4.30	45.50	290	—

TABLE B-9 (Cont.)

Substance	State[b]	Inflammability limit (vol %)[a]		Ignition temperature (°C)	
		Lower	Upper	In air	In O_2
Kerosene	l	—	—	254–293	—
Methane	g	5.00	15.00	650–750	556–700
Methyl alcohol	g	6.72	36.50	470	—
Propane	g	12.12	9.35	504	490–570
Propylene	g	2.00	11.10	558	—
Sulfur	s	—	—	243	—
Toluene	g	1.27	6.75	810	552
Xylene	g	1.00	6.00	—	—

[a]At 20°C, 1 atm in air.
[b]g = gas, l = liquid, s = solid.

PYROMETRIC CONE EQUIVALENT (PCE)

Pyrometric Cone Equivalent is defined as "the number of that standard cone whose tip would touch the supporting plaque simultaneously with that of a cone of the material being investigated, when tested in accordance with the Standard Method of Test for Pyrometric Cone Equivalent (PCE) of refractory materials (ASTM Designation C 24) of the American Society for Testing Materials" (June 1955).

The temperatures given in Table C-1 apply only to the small (1 1/8-in.) PCE cones when heated in an oxidizing atmosphere at the prescribed rate of 150° per hour.

The PCE test is not to be regarded as an accurate measurement of temperature, but rather as "a comparison of thermal behavior in terms of standard cones." While the test does not provide an index of the behavior of a refractory in service, it is useful in evaluating the relative refractoriness of fireclay and some classes of high-alumina refractory materials. It is not applicable to basic and some other refractories.

TABLE C-1

Pyrometric Cone Equivalent (PCE): Cone Temperatures
(to the nearest 5°)[a]

Cone no.	End points	
	°F	°C
12	2440	1335
12–13	2450	1345
13	2460	1350
13–14	2505	1375
14	2550	1400
14–15	2580	1415
15	2605	1430
15–16	2660	1460
16	2715	1490
16–17	2735	1500
17	2755	1510
17–18	2765	1515
18	2770	1520
18–19	2790	1530
19	2805	1540
19–20	2825	1555
20	2845	1565
20–23	2885	1585
23	2920	1605
23–26	2935	1615
26	2950	1620
26–27	2965	1630
27	2985	1640
27–29	3000	1650
29	3020	1660

TABLE C-1 (Cont.)

Cone no.	End points	
	°F	°C
29-31	3040	1670
31	3060	1685
31-31 1/2	3075	1690
31 1/2	3090	1700
31 1/2-32	3105	1710
32	3125	1715
32-32 1/2	3130	1720
32 1/2	3135	1725
32 1/2-33	3150	1735
33	3170	1745
33-34	3185	1755
34	3205	1765
34-35	3225	1775
35	3245	1785
35-36	3260	1795
36	3280	1805
36-37	3295	1810
37	3310	1820
38	3360	1850*
39	3390	1865*
40	3425	1885*
41	3580	1970*
42	3660	2015*

[a](*) Based on the work of Fairchild and Peters, Characteristics of Pyrometric Cones, J. Am. Cer. Soc., 9:700 (1926). Cones Numbers 39 through 42 were heated at a rate of 600°C/hr.

NOTES AND REFERENCES

1. See ASTM fuel testing standards.

2. H. H. Lowry (ed.), "Chemistry of Coal Utilization," Chap. 4 (by W. S. Selvig and F. H. Gibson), John Wiley and Sons, New York, 1945.

3. D. L. Wilson, Prediction of Heat of Combustion of Solid Wastes from Ultimate Analyses, Environ. Sci. Technol., $\underline{6}$:13 (1972).

4. J. H. Perry (ed.), "Chemical Engineers Handbook," 3rd ed., McGraw-Hill, New York, 1950.

5. ASHRE Handbook, ASHRE, New York, 1974.

6. The heating value of amorphous carbon varies somewhat with the source. The value given is an average.

7. F. D. Rossini, J. Res. Nat. Bur. Stand., $\underline{22}$:407 (1939).

8. F. R. Bichowsky and F. D. Rossini, "The Thermochemistry of Chemical Substances," Reinhold, New York, 1936.

9. International Critical Tables, Vol. V, p. 162, McGraw-Hill, New York, 1929.

10. S. Arrhenius, Z. Phys. Chem., $\underline{4}$:226 (1889).

11. H. C. Hottel, G. C. Williams, N. M. Nerheim, and G. Schneider, Combustion of Carbon Monoxide and Propane, 10th Symposium (Int'l.) on Combustion, Combustion Institute, Pittsburgh, Pa., 1965, pp. 111-121.

11a. A. C. Morgan, Combustion of Methane in a Jet Mixed Reactor, D.Sci. thesis, M.I.T., Cambridge, Massachusetts, 1967.

12. L. Shindman, "Gaseous Fuels," American Gas Assn., New York, 1948.

13. M. A. Field, D. W. Gill, B. B. Morgan, and P. E. W. Hawksley, "Combustion of Pulverized Coal," British Coal Utilization Research Assn., Leatherhead, Surrey, England, 1967.

14. G. Guoy, Ann. Chim. Phys., $\underline{18}$:27 (1879).

15. R. M. Fristrom and A. Westenberg, Applied Physics Lab/Johns Hopkins University CM 875 (1955).

16. H. C. Hottel, Burning in Laminar and Turbulent Fuel Jets, 4th Symposium (Int'l.) Combustion, pp. 97-113, Williams and Wilkins, Baltimore, 1953.

17. K. Akita and T. Yumoto, 10th Symposium (Int'l.) on Combustion, Combustion Institute, Pittsburgh, Pa., 1967, p. 943.

18. V. I. Blinov and G. N. Khudyakov, Diffusive Burning of Liquids, Izv. Akad. Nauk SSSR, Ser. Fiz., 1961.

19. J. DeRis and L. Orloff, A Dimensionless Correlation of Pool Burning Rates, Combust. Flame, 18:381-388 (1972).

20. F. R. Steward, Prediction of the Height of Turbulent Diffusion Buoyant Flames, Combust. Sci. Technol., 2:203-202 (1970).

20a. A. Williams, The Mechanism of Combustion of Droplets and Sprays of Liquid Fuels, Oxidation Combust. Rev., 3:1-45 (1968).

21. W. R. Niessen, S. H. Chansky, E. L. Field, A. N. Dimitriou, C. R. La Mantia, R. E. Zinn, T. J. Lamb, and A. S. Sarofim, "Systems Study of Air Pollution from Municipal Incineration," NAPCA, U.S. DHEW, Contract CPA-22-69-23, March, 1970.

22. E. R. Kaiser and S. B. Friedman, Paper presented at 60th Annual Meeting AIChE, November 1968.

23. D. A. Hoffman and R. A. Fritz, Environ. Sci. Technol., 2:1023 (1968).

24. M. A. Kanury, Thermal Decomposition Kinetics of Wood Pyrolysis, Combust. Flame, 18:75-83 (1972).

25. M. A. Kanury, Ph.D. thesis, Univ. Minn., Minneapolis, 1969.

26. K. Akita, Rep. Fire Res. Inst. Japan, 9:1 (1959); and K. Akita and J. M. Kase, Polymer Sci. (A-1), 5:833 (1967).

27. S. L. Madorsky, Thermal Degradation of Organic Polymers, Polymer Rev., 7:238 (1964).

28. U. K. Shivadev and H. W. Emmons, Thermal Degradation and Spontaneous Ignition of Paper Sheets in Air by Irradiation, Combust. Flame, 22:223-236 (1974).

29. P. Nicholls, "Underfeed Combustion, Effect of Preheat, and Distribution of Ash in Fuel Beds," U.S. Bureau of Mines, Bulletin 378 (1934).

30. R. H. Stevens et al., "Incinerator Overfire Mixing Study—Demonstration of Overfire Jet Mixing," OAP, U.S. EPA Contract 68020204 (1974).

31. E. R. Kaiser, Personal communication to W. R. Niessen, C. M. Mohr, and A. F. Sarofim (1970).

32. L. Schiller and A. Naumann, Z. Ver. Dent. Ing., 77:318 (1933).

33. E. J. Schulz, "Dust Emissions from the Hamilton Avenue Incinerator," Battelle Memorial Institute, 11 Sept. 1964.

34. A. B. Walker and F. W. Schmitz, Characteristics of Furnace Emissions from Large, Mechanically Stoked Municipal Incinerators, Proc. 1964 Nat. Incin. Conf., ASME, New York, 1964, pp. 64-73.

35. H. Eberhardt and W. Mayer, Experiences with Refuse Incinerators in Europe, Proc. 1968 Nat. Incin. Conf., ASME, New York, 1968, pp. 142-153.

36. F. Nowak, Erfahrungen an der Müllnerbrennungs—anlage Stuttgart, Brennst-Wärme-Kraft, 19:71-76 (1967).

37. R. L. Stenburg, R. R. Horsley, R. A. Herrick, and A. H. Rose, Jr., Effects of Design and Fuel Moisture on Incinerator Effluents, JAPCA, 10:114-120 (1966).

38. E. R. Kaiser, C. O. Zeit, and J. B. McCaffery, Municipal Incinerator Refuse and Residues, Proc. 1968 Nat. Incin. Conf., ASME, New York, 1968, p. 142.

39. E. R. Kaiser, Chemical Analysis of Refuse Components, Proc. 1966 Nat. Incin. Conf., ASME, New York, 1966, p. 84.

40. E. R. Kaiser, Refuse Composition and Flue-Gas Analysis from Municipal Incinerators, Proc. 1964 Nat. Incin. Conf., ASME, New York, 1964, pp. 35-51.

41. R. L. Stenberg, R. P. Hangebrauck, D. J. Von Lehmden, and A. H. Rose, Jr., Field Evaluation of Combustion Air Effects on Atmospheric Emission from Municipal Incinerators, JAPCA, 12:83-89 (1962).

42. R. G. Engdahl, Combustion in Furnaces, Incinerators and Open Fires, in "Air Pollution," Vol. III (A. C. Stern, ed.), 2nd ed., Reinhold, New York, 1968, pp. 4-54.

43. A. H. Rose, Jr. and H. R. Crabaugh, Research Findings in Standards of Incinerator Design, in "Air Pollution" (Mallette, ed.), 1st ed., Reinhold, New York, 1955.

44. P. H. Woodruff and G. P. Larson, Combustion Profile of a Grate-Rotary Kiln Incinerator, Proc. 1968 Nat. Incin. Conf., ASME, New York, 1968, pp. 327-336.

45. H. G. Meissner, The Effect of Furnace Design and Operation on Air Pollution from Incinerators, Proc. 1964 Nat. Incin. Conf., ASME, New York, 1964, pp. 126-127.

46. F. R. Rehm, Incinerator Testing and Test Results, JAPCA, 6:199-204 (1957).

47. A. H. Rose, Jr., R. L. Stenburg, H. Corn, R. R. Horsley, D. R. Allen, and P. W. Kolp, Air Pollution Effects of Incinerator Firing Practices and Combustion Air Distribution, JAPCA, 8:297-309 (1959).

48. E. S. Grohse and L. E. Saline, Atmospheric Pollution: the Role Played by Combustion Processes, JAPCA, 8:255-267 (1958).

49. H. F. Johnstone, Univ. Ill. Eng. Exp. Station Bull., 228:221 (1931).

50. B. Zeldovitch, P. Sadovnikov, and D. Frank-Kamenetski, "Oxidation of Nitrogen in Combustion," Academy of Sciences (USSR), Inst. of Chem. Physics, Moscow-Leningrad, 1947.

51. A. A. Westenberg, Combust. Sci. Technol., 4:59 (1971).

52. "Air Quality and Stationary Source Emission Control," Nat. Acad. Sci. Serial 94-4, March 1975, pp. 817-884.

53. C. T. Bowman, Kinetics of Pollutant Formation and Destruction in Combustion, Prog. Energy Comb. Sci., 1:33-45 (1975).

54. R. L. Stenburg, R. P. Hangerbrauck, D. J. Von Lehmden, and A. H. Rose, Jr., Effects of High Volatile Fuel on Incinerator Effluents, JAPCA, 11:376-383 (1961).

55. "Modern Refractory Practice," Harbison-Walker Refractories Company, William Feather Co., Cleveland, Ohio, 1961.

56. H. H. Krause, D. A. Vaughan, and W. K. Boyd, Corrosion and Deposits from Combustion of Solid Waste, Part IV. Combined Firing of Refuse and Coal, Proc. ASME Winter Annual Meeting, Houston, Texas, 1975.

57. H. H. Krause, D. A. Vaughan, and P. D. Miller, Corrosion and Deposits from Combustion of Solid Waste, J. Eng. Power, trans. ASME, Series A, 95:45-52 (1973).

58. H. H. Krause, D. A. Vaughan, and P. D. Miller, Corrosion and Deposits from Combustion of Solid Waste. Part 2. Chloride Effects on Boiler Tube and Scrubber Metals, J. Eng. Power, trans. ASME, Series A, 96:216-222 (1974).

59. H. H. Krause, D. A. Vaughan, and W. K. Boyd, Corrosion and Deposits from Combustion of Solid Waste. Part 3. Effects of Sulfur on Boiler Tube Metals, ASME Winter Annual Meeting, 1974.

60. P. D. Miller and H. H. Krause, Corrosion of Carbon and Stainless Steels in Flue Gases from Municipal Incinerators, Proc. 1972 ASME Nat. Incin. Conf., ASME, New York, 1972, p. 300.

61. P. D. Miller and H. H. Krause, Factors Influencing the Corrosion of Boiler Steels in Municipal Incinerators, Corrosion, 27:31-45 (1971).

62. F. Nowak, Considerations in the Construction of Large Refuse Incinerators, Proc. 1970 ASME Nat. Incin. Conf., ASME, New York, 1970, pp. 86-92.

63. H. C. Hottel and A. F. Sarofim, "Radiative Transfer," McGraw-Hill, New York, 1967.

64. H. S. Carslaw and J. C. Jaeger, "Conduction of Heat in Solids," Oxford Univ. Press, Oxford, England, 1959.

65. W. R. Niessen, M. C. Mohr, R. W. Moore, and A. F. Sarofim, "Incinerator Overfire Mixing Study," CSD, OAP, U.S. EPA Contract EH SD 71-6, February, 1972.

66. A. C. Stern, "Abating the Smoke Nuisance," Mechanical Eng., 54:267–268 (1932).

67. J. A. Switzer, The Economy of Smoke Prevention, Eng. Mag., 406–412 (1910).

68. H. Kreisinger, C. E. Augustine, and F. K. Ovitz, Combustion of Coal and Design of Furnaces, U.S. Bur. Mines Bull. 135, 1917.

69. L. P. Breckenridge, Study of Four Hundred Steaming Tests, U.S. Geol. Survey Bull. 325:171–178 (1907).

70. A. E. Grunert, Increasing the Oxygen Supply Over the Fire, Power, 130–131 (1927).

71. M. K. Drewry, Overfire Air Injection with Underfeed Stokers, Power, 446–447 (1926).

72. A. R. Mayer, Effect of Secondary Air in the Traveling Grate Stoker Furnace, Z. Bayr. Rev. Verein, 42 (1938).

73. R. F. Davis, The Mechanics of Flame and Air Jets, Proc. Inst. Mech. Eng., 137:11–72 (1937).

74. E. W. Robey and W. F. Harlow, Heat Liberation and Transmission in Large Steam-Generating Plant, Proc. Inst. Mech. Eng., 125:201 (1933).

75. W. Gumz, Overfire Air Jets in European Practice, Combustion, 22:39–48 (1951).

76. "The Reduction of Smoke from Merchant Ships," Fuel Research Technical Paper No. 54 issued by the Department of Scientific and Industrial Research, London, England, 1947.

77. J. A. Switzer, Smoke Prevention with Steam Jets, Power, 75–78 (1912).

78. L. N. Rowley and J. C. McCabe, Cut Smoke by Proper Jet Application, Power, 70–73 (1948).

79. F. B. Tatom, Sc.D. thesis, Georgia Institute of Technology, Atlanta, Georgia, 1971.

80. R. F. Davis, The Mechanics of Flame and Air Jets, Proc. Inst. Mech. Eng., 137:11–72 (1937).

81. Y. V. Ivanov, "Effective Combustion of Overfire Fuel Gases in Furnaces," Estonian State Pub. House, Tallin, USSR, 1959.

82. E. R. Kaiser and J. B. McCaffery, "Overfire Air Jets for Incinerator Smoke Control," Paper 69-225 presented at Annual Meeting APCA, June 26, 1969.

83. G. N. Abramovich, "The Theory of Turbulent Jets" (L. H. Schindel, ed.), M.I.T. Press, Cambridge, Mass., 1963. (Transl. by Scripta Technica.)

84. S. Corrsin, NACA, Wartime Reports No. ACR 3L23.

85. F. P. Ricou and D. B. Spaulding, J. Fluid Mech., 11:21-32 (1961).

86. A. N. Syrkin and D. N. Lyakhovskiy, "Aerodynamics of an Elementary Flame," Sooshch. Isentr. Nauchn-Issled. Kotloturvinnyi Inst., 1936.

87. M. A. Patrick, Experimental Investigation of the Mixing and Penetration of a Round Turbulent Jet Injected Perpendicularly into a Transverse Stream, Trans. Inst. Chem. Eng., 45:T-16 to T-31 (1967).

88. G. S. Shandorov, Flow From a Channel into Stationary and Moving Media, Zh. Tekhn. Fiz., 37:1 (1957).

89. J. F. Keffer and W. D. Barnes, J. Inst. Fuel, 15:481-496 (1963).

90. W. Tollmien, Berechnung Turbulenter Ausbreitungsvorange, Z. fur Angewandte Mathematik Mechanik, 6:468 (1926).

91. "Layout and Application of Overfire Jets for Smoke Control in Coal Fired Furnaces," National Coal Association, Washington, DC, Section F-3, Fuel Engineering Data, Dec. 1962.

92. M. W. Thring, "The Science of Flames and Furnaces," Chapman and Hall, London, 1962, p. 164.

93. N. Syred and J. M. Beer, Combustion in Swirling Flows: A Review, Combust. Flame, 23:143-201 (1974).

94. J. M. Beér and W. Leuckel, "Turbulent Flow in Rotating Flow Systems," Paper No. 7, North American Fuels Conference, Ottawa, Canada, May 1970 (Canadian Combustion Institute, ASME, and the Institute of Fuel).

95. N. M. Kerr and D. Fraser, J. Inst. Fuel, 38:519-526 (1965).

96. N. Syred and J. M. Beér, Proc. 14th Int'l. Symposium on Combustion, Combustion Institute, Pittsburgh, Pa., 1973, p. 523.

97. S. A. Tager, Therm. Eng., 18:120 (1972).

98. Y. V. Troyankin and E. D. Balnev, Therm. Eng., 16:45 (1969).

99. G. G. Bafuwa and N. R. L. Maccallum, "Turbulent Swirling Flames Issuing from Vane Swirlers," Paper presented to the 18th Meeting of the Aerodynamics Panel, International Flame Research Foundation, Paris, September, 1970.

100. S. A. Beltagui and N. R. L. Maccallum, "Aerodynamics of Swirling Flames—Vane Generated Type," Proc. Comb. Inst., European Symposium, Sheffield University, 1973, p. 599.

101. J. M. Beér and K. B. Lee, Proc. 10th Int'l. Symposium on Combustion, Combustion Institute, Pittsburgh, Pa., 1965, p. 1187.

102. P. F. Drake and E. F. Hubbard, J. Inst. Fuel, 39:98 (1966).

103. M. W. Thring and M. P. Newby, Proc. 4th Int'l. Symposium on Combustion, Combustion Institute, Pittsburgh, Pa., 1953, pp. 739-796.

104. A. Craya and R. Curtet, C.R. Acad. Sci., Paris, 241:621-622 (1955).

105. M. Barchilon and R. Curtet, J. Basic. Eng., 86D:777-787 (1964).

106. P. G. Hill, J. Fluid Mech., 22:161-186 (1965).

107. N. S. Ulmer, "Physical and Chemical Parameters and Methods for Solid Waste Characterization," Open File Progress Report, RS-03-68-17, July 1967-July 1969, OSWMP, 1970.

108. W. R. Niessen, Estimation of Solid Waste Production Rates; and Refuse Characteristics, in "Handbook of Solid Waste Management" (D. G. Wilson, ed.), Van Nostrand-Reinhold, New York, 1977, pp. 10-62, 544-574.

109. W. R. Niessen and S. H. Chansky, The Nature of Refuse, Proc. 1970 ASME Incin. Conf., ASME, New York, 1970, p. 1.

110. W. R. Niessen and A. F. Alsobrook, Municipal and Industrial Refuse: Composition and Rates, Proc. 1972 ASME Incin. Conf., ASME, New York, 1972, p. 319.

111. Anon., World Survey Finds Less Organic Matter, Refuse Removal J., 10:26 (1967).

112. F. H. Miller, "Conversion of Organic Solid Wastes into Yeast; An Economic Evaluation," PHS Pub. 1901, 1969, p. 21.

113. K. Matoumoto, R. Asakata, and T. Kawashima, The Practice of Refuse Incineration in Japan Burning Refuse with High Moisture Content and Low Calorific Value, Proc. 1968 ASME Incin. Conf., ASME, New York, 1968, p. 180.

114. "Solid Wastes Study and Planning Grant, Jefferson County, Ky," U.S. Dept. HEW, 1967.

115. "The Solid Waste Disposal Study, Genesee County, Mich.," U.S. Dept. HEW, 1968.

116. E. R. Kaiser and A. A. Caroti, "Municipal Incineration of Refuse with 2% and 4% Additions of Four Plastics," Report to Society of the Plastics Industry, 1971.

117. E. R. Kaiser, Incineration of Bulky Refuse III, Proc. 1972 ASME Incin. Conf., ASME, New York, 1972, p. 72.

118. E. R. Kaiser, Incineration of Bulky Refuse, Proc. 1966 ASME Incin. Conf., ASME, New York, 1966, p. 39.

119. "Omaha-Council Bluffs Solid Waste Management Plan-Status Report 1969," EPA, OSWMP Pub. SW-3t s g (1971).

120. "High Pressure Compaction and Baling of Solid Waste," EPA, OSWMP, Pub. SW-32 d (1972).

121. E. R. Kaiser, The Incineration of Bulky Refuse II, Proc. 1968 ASME Incin. Conf., ASME, New York, 1968, p. 129.

122. G. R. Davidson, Jr., "A Study of Residential Solid Waste Generated in Low-Income Areas, " EPA, OSWMP Pub. SW-83 t s (1972).

123. "Industrial Solid Waste Survey, Oregon: 1970," Oregon State Board of Health, Solid Waste Section (1971).

124. W. T. Ingram and F. P. Francia, "Quad City Solid Wastes Project— Interim Report, " Paper prepared for U.S. Dept. HEW by Quad City Solid Wastes Committee Staff (1968).

125. J. J. Baffo and N. Bartilucci, "Bulky and Demolition Wastes," Report to the City of New York, John Baffo Consulting Engineers, New York, 1968.

126. "District of Columbia Solid Waste Management Plan-Status Report 1970," EPA, OSWMP Pub. SW-4 t s g (1971).

127. R. C. Corey (ed.), "Principles and Practices of Incineration," Wiley-Interscience, New York, 1969, p. 83.

128. N. A. Lange (ed.), "Langes' Handbook of Chemistry," 10th ed. (revised), McGraw-Hill, New York, p. 1577, 1969.

129. "Municipal Refuse Disposal," Institute for Solid Wastes, Am. Publ. Works Assn., p. 179, 1970.

130. F. E. Wisely and H. B. Hinchman, Refuse as a Supplementary Fuel, Proc. 3rd Ann. Env. Eng. Sci. Conf., Louisville, Ky., March 5-6, 1973.

131. N. W. Wagner, Discussions of Papers, ASME Incinerator Conference, ASME, New York, 1968, pp. 5, 6.

132. J. A. Duffie and W. R. Marshall, Jr., Factors Influencing the Properties of Spray Dried Materials, Chem. Eng. Progr., 49:417-423 (1953).

133. F. W. Rohr, Suppression of a Steam Plume from Incinerator Stacks, Proc. 1968 ASME Incin. Conf., ASME, New York, 1968, pp. 216-224.

134. J. A. Fife and R. H. Boyer, What Price Incineration, Air Pollution Control, Proc. 1966 ASME Incin. Conf., ASME, New York, 1966, pp. 89-96.

135. R. J. Schoenberger, Studies of Incinerator Operation, Proc. 1972 ASME Incin. Conf., ASME, New York, 1972, p. 15.

136. W. C. Achinger and L. E. Daniels, An Evaluation of Seven Incinerators, Proc. 1970 ASME Incin. Conf., ASME, New York, 1970, p. 32.

137. "Incinerator Standards," Incinerator Institute of America, New York, Nov. 1968.

138. J. E. Williamson, R. J. MacKnight, and R. L. Chass, "Multiple-Chamber Incinerator Design Standards," Los Angeles APCD, October, 1960.

139. J. A. Danielson (ed.), "Air Pollution Engineering Manual," 2nd ed., U.S.E.P.A., p. 434 et seq., 1973.

140. E. S. Monroe, "New Developments in Industrial Incineration," Proc. 1966 ASME Incin. Conf., ASME, New York, 1966, pp. 226-230.

141. Bayard, Chem. Metallurgical Eng., 52:100-102 (1945).

142. E. M. Smith and A. R. Daly, "The Past, Present and Future Prospects of Burning Municipal Sewage Sludge along with Mixed Municipal Refuse," Second National Conference on Municipal Sludge Management and Disposal, Anaheim, Calif., 1975.

143. J. Defeche, "Combined Disposal of Refuse and Sludges: Technical and Economic Considerations," 1st International Congress on Solid Wastes Disposal and Public Cleansing ISWA, Praha, 1972, Thema V, June/July, 1972, pp. 3-39.

144. F. Rub, Possibilities and Examples of a Combined Incineration of Refuse and Waste Water Sludge, Wasser Luft Betrieb, 14:484-488 (1970).

145. Anon., The Reigate Incinerator, Surveyor (London), 138:34-35 (1971).

146. Anon., Solid Waste and Sludge = Energy Self-Sufficiency, Resource Recovery Energy Review, 2:16, 17 (1975).

147. E. M. Smith, A. R. Daly, and W. R. Niessen, Co-incineration of Municipal Sewage Sludge and Municipal Refuse, EPA, OWP (1976).

148. M. Leva, "Fluidization," McGraw-Hill, New York, 1959.

149. D. Kunnii and O. Levenspeil, "Fluidization Engineering," Wiley, New York, 1969.

150. Vanecek, Markvat, and Drbohlav, "Fluidized Bed Drying," Leonard Hill, London, 1966.

151. D. L. McGill and E. M. Smith, Fluidized Bed Disposal of Secondary Sludge High in Inorganic Salts, Proc. 1970 ASME Incin. Conf., ASME, New York, 1970, pp. 79-85.

152. "Control Techniques for Hydrocarbon and Organic Solvent Emissions from Stationary Sources," USDHEW, PHS, NAPCA Publication No. AP-68, March, 1970.

153. "Control Techniques for Particulate Air Pollutants," USDHEW, PHS, NAPCA Publication No. Ap-51, January, 1969.

154. W. R. Niessen and A. F. Sarofim, "Air Pollution Control for Incinerators," presented at the National Industrial Solid Wastes Conference, Univ. Houston, Houston, Texas, March 25, 1970.

155. J. P. Longwell and M. A. Weiss, Ind. Eng. Chem., 47:1634 (1955).

156. A. R. Mayer, Untersuchung uber Zweitluftfuhr in Wandercostfeurungen, Feuerungstechnik, 26:201-210 (1938).

157. G. H. N. Chamberlain and A. D. Walsh, Processes in the Vapour Phase Oxidation of Ether II, 3rd Symposium (Int'l.) on Combustion, pp. 368-374, Combustion Institute, Pittsburgh, Pa., 1949.

158. M. Weintraub et al., "Experimental Studies of Incineration in a Cylindrical Combustion Chamber," Bureau of Mines, RI 6908, 1967.

159. G. L. Brewer, Fume Incineration, Chem. Eng., 75:160-165 (1968).

160. S. Z. Roginskii, "Problemy Kinetiki i Kataliza," Vol. IV, 1940, p. 187.

161. C. Wagner, J. Chem. Phys., 18:69 (1950).

162. R. M. Dell, F. S. Stone, and P. F. Tiley, Trans. Faraday Soc., 49:201 (1953).

163. K. Hanffc, R. Gland, and H. J. Engell, Z. Physik Chem., Leipzig, 201:223 (1952).

164. D. A. Dowden, J. Chem. Soc., 242 (1952).

165. W. Sachtler and J. Fahrenfort, Actes 2e Cong. Intern. Catalyses, Paris (1960), pp. 1499, Editions Technip, Paris, 1961.

166. R. B. Anderson, K. C. Stein, J. J. Feenan, and L. J. E. Hofer, I/E.C. 53:809-812 (1961).

167. L. S. Caretto and K. Nobe, I/E.C. (Process Des. Dev.), 5:217 (1966).

168. G. Vollheim, Twelfth International Symposium on Combustion, University of Poitiers, Poitiers, France, July 14-20, Combustion Institute, Pittsburgh, Pa., 1968, p. 653.

169. E. W. Thiele, I/E.C., 31:916 (1934).

170. C. N. Satterfield and T. K. Sherwood, "The Role of Diffusion in Catalysis," Addison-Wesley, Reading, Massachusetts, 1963.

171. T. M. Hein, Ann. N.Y. Acad. Sci., 116:656-662 (1964).

172. J. L. Donahue, J. Air Pollution Control Assoc., 8:209 (1958).

173. W. R. Niessen, S. E. Van Vliet, R. Grzywinski, and M. E. Fiore, "Solid Waste Management/Resources Recovery Plan, Richmond Metropolitan Area," Roy F. Weston, West Chester, Pa., 1975.

174. Stanley Consultants, Inc., "Sludge Handling and Disposal, Phase I, State-of-the-Art, Report to Metropolitan Sewer Board of the Twin Cities Area," 1972.

175. O. E. Albertson, "Low Cost Combustion of Sewage Sludges," Dorr-Oliver Inc., Technical Pre-print No. 600-P.

176. G. J. Ducar and P. Levin, "Mathematical Model of Sewage Sludge Fluidized Bed Incinerator Capacities and Costs," FWQCA Report, TWRC-10, 1969.

177. A. H. Manchester, "Comparison Between Fluid Bed Incineration and Multiple Hearth Incineration," private communication.

178. G. G. Copeland and I. G. Lutes, Fluidized Bed Combustion of Sewage Sludge, Eng. Digest (April 1973).

179. M. Allenspach, "Cost Analysis of Fume Incinerators," 62nd Annual Air Pollution Conference, APCA.

Note: Page numbers referencing tables are underlined.